P9-EGL-173

THE ESSENTIAL GOURD

THE ESSENTIAL GOURD
Art and History in Northeastern Nigeria

Marla C. Berns
Barbara Rubin Hudson

MUSEUM OF CULTURAL HISTORY
University of California, Los Angeles

This publication and the exhibition associated with it were supported by a grant from the National Endowment for the Humanities, Washington, D.C., a Federal Agency created by Act of Congress, 1965. Further support was provided by the Ahmanson Foundation and Manus, the support group of the UCLA Museum of Cultural History.

Exhibition presented at:

Wight Art Gallery, University of California, Los Angeles
March 25–May 18, 1986
Honolulu Academy of Art, Honolulu
March 25–May 3, 1987
The Center for African Art, New York City
June 10–September 10, 1987
National Museum of African Art, Washington, D.C.
January 26–May 22, 1988

Published in conjunction with the Seventh Triennial Symposium on African Art, University of California, Los Angeles, April 2–6, 1986, hosted by the UCLA Museum of Cultural History.

Cover. Decorated gourds from Northeastern Nigeria.
UCLA Museum of Cultural History collection.
Frontispiece. Pidlimndi woman carrying pyro-engraved bowl
wrapped with homespun cloth; Mbagu. 1971.

Museum of Cultural History
University of California, Los Angeles
405 Hilgard Avenue
Los Angeles, California 90024

ISBN 0-930741-08-0 (softcover)
ISBN 0-930741-07-2 (casebound)
Library of Congress Catalog Card Number: 85-48263

Second Printing, 1992

Printed in the United States of America

CONTENTS

PREFACE

This publication includes material drawn from two independent field research projects in northern Nigeria separated by a ten-year interval. The majority of the gourds illustrated and discussed were field-collected by Barbara Rubin Hudson between 1969–1971. She conducted research among groups who still actively practiced this art and focused specifically on techniques, styles of decoration, and meanings of design systems. Her intensive concentration on this single category of production has yielded precise information about artists, innovations, and relationships across ethnic boundaries. In 1980–1982, Marla Berns worked among many of the same groups; however, calabash decoration was only one component of a larger project aimed at documenting the entire range of arts produced by the little-known groups living in the region. Linguistic, ethnographic, and historical data were used to correlate the distribution of these arts within the Lower Gongola Valley to a reconstruction of the ethnohistory of their producers. A number of gourds also were collected during Berns' field investigations.

This project had its genesis in the donation by Barbara Rubin Hudson of 214 decorated gourds to the Museum of Cultural History. Dr. Hudson's generosity to the Museum and her contributions to this catalogue are greatly appreciated. A second donation of 61 gourds from the same area by Linda and Bruce Friedman provided an excellent complement to the initial acquisition.

Marla Berns as principal author of this book and curator of the associated exhibition has brought three years of fieldwork and a warm affection for the people of Northern Nigeria to this project. Through her efforts what was initially envisioned as a modest catalogue has grown into a full-scale book detailing the richness of what on the surface might appear to be a humble art form. It has been a pleasure for the Museum staff to work closely with Dr. Berns on all phases of the publication and exhibition. She has our sincerest thanks.

The quarterly, *African Arts*, published by the UCLA African Studies Center has been continually supportive of Museum of Cultural History projects. They generously featured a preview chapter of this volume and provided a number of color separations used in its production. We would like to thank John Povey, editor, Alice McGaughey, art director, and Amy E. Futa, executive editor, for valued help. UCLA Chancellor Charles E. Young and Vice Chancellor Elwin V. Svenson have also provided ongoing institutional backing for all Museum programs.

This volume is being published in conjunction with the Seventh Triennial Symposium on African Art hosted by the Museum of Cultural History. We thank the numerous co-sponsors of the Triennial for ensuring the success of the symposium.

As the Museum of Cultural History prepares for its move into a new Museum facility in 1988, "The Essential Gourd" marks the last of the Museum's traveling exhibitions to be mounted in the Frederick S. Wight Art Gallery. We again thank the staff of the Gallery, directed by Dr. Edith Tonelli, for their full collaboration in this and past projects. The Museum's international reputation has largely been built on traveling exhibitions initially mounted in the Wight Art Gallery. The efforts of all the Gallery staff members over the past years are sincerely appreciated.

The exhibition has benefited from the support of numerous individuals. Al and Edna Heer from the Pumpkin Farm and Gourd Place in Paso Robles, California, provided crucial display materials and information on the cultivation and cleaning of gourds. Robin Wells of the Treganza Museum, San Francisco State University, facilitated the loan of gourds field-collected by David Ames. Russell and Maxine Schuh also provided important Hausa gourds.

As the exhibition evolved, it was expanded to include decorated gourds and objects made with gourds from the rest of Africa. We thank Herbert M. Cole, Ernie Wolfe III, Jim and Jean Willis, Barry and Jill Kitnick, and Robert and Helen Kuhn, who lent key objects for this section. René and Stephanie Bravmann donated a Dodo gourd mask from Burkina Faso, and Geoffrey Conrad of the William Hammond Mathers Museum, Indiana University, approved the loan of a Songye gourd mask from Zaire collected by Alan Merriam. Critical research assistance was provided by Rachel Hoffman who also generated most of the labels in this area.

The life of the northeastern Nigerian section of this exhibition has been extended with a national tour. For their participation in this project, we thank George R. Ellis of the Honolulu Academy of Art, Susan Vogel of the Center for African Art, and Sylvia Williams and Roy Seiber of the National Museum of African Art.

Funding for this volume was provided by the National Endowment for the Humanities. The assistance of Sally Yerkovich from the Endowment is gratefully acknowledged. Additional funding was contributed by the Ahmanson Foundation and by Manus, the support group of the Museum of Cultural History. The Foundation and the members of Manus are directly responsible for the growth of the Museum's publication and exhibition programs over the past five years. They have our enthusiastic gratitude.

Christopher B. Donnan, *Director*
Doran H. Ross, *Associate Director*

ACKNOWLEDGMENTS

This work is the result of a longstanding interest in the Nigerian art of decorating gourds. Its genesis dates to the beginning of the dry season, 1969, when Barbara Rubin Hudson observed the beauty and ubiquity of this artistic tradition:

> I recall that dry season vividly, because I found myself alone and in charge of two small children, a dog, several chickens, a houseboy, and a cook, all of us inhabitants of a bush rest house in Biu, Borno Province, Northeast State, Nigeria, as it was called then. One could hardly fail to notice the daily parade on the road in front of the house. Women on their way to market, women going to and from their farms, women with their babies on their backs, women everywhere with the most remarkable, most exuberantly decorated gourds carried on their heads, against their shoulders, and over their babies' heads. Soon, I was standing at the roadside asking questions: where were the gourds from, who was decorating them, did the designs have names or meanings? The answers only provoked more questions: always, the women were telling me about other people, other groups, just down the road, or over the hill, or across the river, or three villages beyond, where gourds were engraved differently or carved or painted or dyed. I knew that this was "field research" only after I abandoned writing on odd scraps of paper and bought several bound notebooks in which I kept a more systematic record of all that I was learning. It was only after returning to the United States that I found this region viewed by scholars as "art poor," perhaps, in retro-

> spect, as good an argument as one might make for going into the field unprepared.

The gourds field-collected by Barbara Rubin Hudson between 1969 and 1971 provided the material foundation for this project. I would like to thank Barbara for personally and professionally encouraging my own investigations in northeastern Nigeria and for contributing many insights and much information to this volume.

My field research in the Gongola-Hawal Valley from September 1980 to June 1982 focused in part on the region's rich traditions of gourd decoration. This work was co-funded by the Fulbright-Hays Doctoral Dissertation Research Abroad Program and by the International Doctoral Research Fellowship Program of the Social Science Research Council. The Edward A. Dickson Fellowship Fund provided support for preparatory research and travel. My thanks are due to all these programs. I also am grateful to the National Commission for Museums and Monuments, Federal Republic of Nigeria and the Center for Nigerian Cultural Studies, Ahmadu Bello University, Zaria, for institutional support during my stay. Both staffs were interested and helpful in a variety of ways. Special thanks, however, are extended to Dr. Ekpo Eyo, Moses Abun, and Mahdi Ahdamu.

Both Barbara and I owe many northeastern Nigerians sincere thanks for their generous assistance and participation. Innumerable artists and patrons shared their knowledge and experiences with warmth and enthusiasm. While we would like to acknowledge them all by name, we have limited our thanks to those artists with whom we developed the closest relationships. Barbara notes the help of Jumai Pitiri Gulcoss, Amsa Galadima, and Aisi Gana from the Tera village of Wuyo, and that of Kalhar and Witebar from the Ga'anda village of Ga'anda. I, too, would like to thank Kalhar and number of Ga'anda artists: Ilimina Kwanta, Ndinuwa Coxita, Fadimatu Ga'anda, and Adio Ga'anda. My deepest appreciation, however, goes to Dije Boka for her unselfish commitment, hospitality, and friendship. I am also grateful to the women of Ga'anda, Gabun, Kwanta, Gangrang, Jebre, Sama, Jaromboyi, Yang, Riji, Kiro, Sheno, Ganjim, Biki, Suktu, Shani, Buma, Kubodeno, Walama, Guyaku, Balhona, Dinga, Linga, Borrong, Talasse, Dela Waja, Swa, and Zambuk for demonstrating their skills and displaying their precious collections. Although I had many capable field assistants, both my and Barbara's work among the Ga'anda was greatly facilitated by Musa Wawu na Hammandikko, our organizer, interpreter, guide, and friend. Additionally, I would like to thank Gerema David for easing the burden of a constantly changing living and working environment, which he did with unfailing resourcefulness and an ever-cheerful demeanor.

Fellow scholars and gourd cognoscenti enthusiastically supplied information, insights, and photographs. Dr. David Ames, Dr. Russell Schuh, and Mr. T.J.H. Chappel served as official consultants. Also helpful were Dr. Charles Heiser, Dr. Jonathan Sauer, Dr. Merrick Posnansky, Dr. Judith Perani, and Ms. Pamela Mauk. Preeminent, however, was the contribution of Dr. Arnold Rubin who as primary consultant read the entire manuscript and offered valuable criticism. He also generously provided information from his field notes, rich with references to the use and meaning of gourds. I am further indebted to Dr. Rubin for being a constant source of inspiration and encouragement during my years as his graduate student; and, above all, for long being a special friend.

Others have contributed to the production of this book and the exhibition. I am grateful to Dr. Thurstan Shaw, Carol Beckwith, Robbie Reid, and the National Commission for Museums and Monuments (Nigeria), for the use of their fine photographs. I am particularly grateful to Ray and Mildred Konan, whose excellent slides have greatly enhanced the exhibition. The Frobenius Institut, Frankfurt, has kindly given permission to reproduce one of Carl Arrien's early watercolors. Nancy Toothman skillfully executed the detailed line drawings. Jeri Bernadette Williams ably served as research assistant in the initial phases of the project. Thanks are also extended to the Melville J. Herskovits Africana Library, Northwestern University, for its research assistance.

The entire staff of the Museum of Cultural History deserves high praise for its commitment to the goals of this project. Under the adept supervision of Christopher Donnan and Doran H. Ross, they have given generously of their time and energy over the past two years. The registration staff, efficiently directed by Sarah Kennington, handled the complex job of accessioning and keeping track of a kaleidoscopic array of gourds. Special compliments go to Jennifer Garman for her hard work during the early stages. Assistant Registrar Verni Greenfield, aided by Intern Paulette Parker, have ably managed a myriad of subsequent details. Robert V. Childs, Collections Manager, and his Assistant, Owen Moore, have also been instrumental to the effective management of a diverse assemblage of objects. Thanks go to Barbara Underwood for her often unseen (but not unsung) administrative contributions. Millicent Besser attended to financial matters promptly and efficiently. Emily M. Woodward wrote an excellent grant application and proved successful in her efforts. Betsy Quick did a fine job of coordinating publicity and developing didactic materials. Benita Johnson conserved a number of gourds, and oversaw their careful handling and installation. Richard Todd took the superb studio photographs and was assisted in the dark room by Ellen Hardy. Darlene Moses Olympius designed the book beautifully and coordinated its production, working under tight deadlines. Irina Averkieff edited the manuscript with special sensitivity and meticulous attention to details of exposition. Susan Swiss provided editorial assistance and spent endless hours at the word processor. Jack Carter and Tom Hartmann designed the exhibition with flair and creativity, while responding perceptively to its curatorial objectives. George Johnson also made valuable contributions to its design in the initial stages. Others at the Museum whose help was indispensable, but whose work takes place behind the scenes, are Betsy Escandor, Phillip M. Douglas, Rashad Raheem, and Jenny Underwood. Most prominent, however, was the contribution of Doran H. Ross, whose commitment to excellence on many levels is infectious and inexhaustible. He has thoughtfully and enthusiastically directed this project since its inception. My sincere thanks are extended to all the Museum staff, who have been friends as well as colleagues for many years.

A final debt of gratitude is owed the close friends and family members who have long supported and encouraged my academic pursuits. This includes the special relationships that developed in Nigeria, which offered practical and emotional sustenance during many long lonely months. It also includes my present colleagues at Neuhart Donges Neuhart, who have been patient with and interested in my preoccupation with gourds. The greatest thanks, however, go to my parents for their love and unconditional endorsement.

Marla C. Berns

Plate 1. Ga'anda Hills with clusters of compounds belonging to the Tumbar family; Ga'anda. December 1980.

Plate 2. Tangale Peak and the Benue-Gongola Valley as seen from the top of Kaltungo Hill. June 1982.

Plate 3. Exhibition of Waja decorated gourds and other examples of material culture; Swa. February 1982.

Plate 4. Prepared gourds—decorated and undecorated—for sale at the daily market in Gombe. July 1982.

Plate 5. Annual fishing festival; Gashua. Fishermen ride on the gourd floats while waiting for a catch; small fish are temporarily stored inside the floats. 1975.

Plate 6. Women selling food in decorated gourd bowls; Biu market. 1970.

Plate 7. Ga'anda decorated gourd bowls (*njoxtitib'a*) displayed in a wickerwork basket (*can'lan'nda*) as they would be during wedding ceremonies. Basket: 49.5 cm. UCLA MCH X85–97; Gourds: UCLA MCH collection.

Plate 8. Pastoral Fulani (Wodabe) display of contents of *kaakel* packs during Worso, the occasion when sublineages gather to celebrate births and marriages of the previous year. 1982.

Plate 9. Interior view of a Tera woman's room (*kɇba*) showing stacked pottery and decorated gourds; Shinga. Note painted display vessels. April 1982.

Plate 11. Mbula cement grave monument and display of the deceased's attributes of social and ritual status; Dilli. In addition to the gourd bottles (*dungala*), the large bundles of skin bags (*ngya luru*) represent the accumulated power and authority of the Ngala healing society. January 1982.

Plate 12.
Yungur woman selling beer at Jangara market. January 1981.

Plate 10.
Waja women using gourd bonnets (*gakire*) to shield their babies from the sun when going to a well to fetch water; Talasse. February 1982.

Plate 13. Dera girls displaying calabash rattles (*lib'e wa*) and other dance attributes used for performances of Bwalin held during Ilela or more recently to enhance collective Menwara festivities; Buma. December 1980.

Plate 14. Three Yungur gourd drums (*dimkedim*), which are played together during Wora second funeral celebrations; Dirma. May 1981.

Plate 15. Hausa gourd rattles (*caki*) being played at the Agricultural Show; Gombi. February 1981.

Plate 16. Waja woman pyro-engraving and showing the broad movements of the technique; Bangu. February 1982.

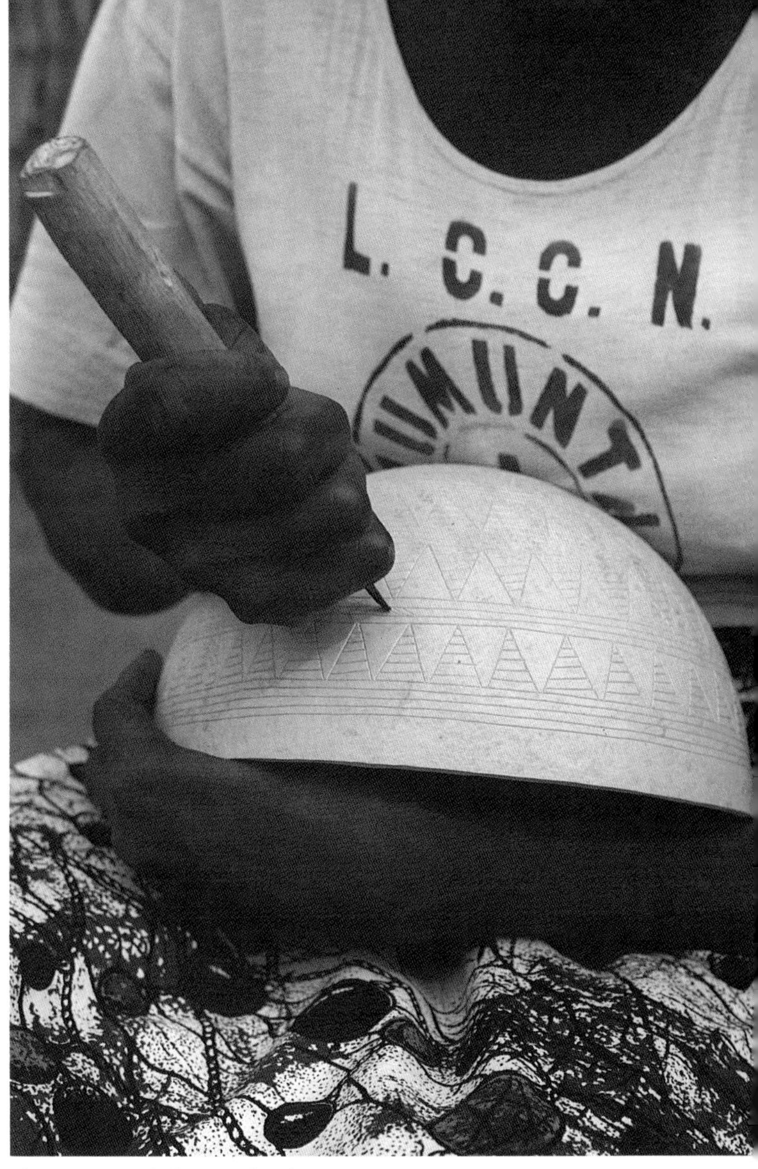

Plate 17. Rahab, a Ga'anda woman, pressure-engraving; Gabun. January 1981.

Plate 20. Stages of interior painted decoration on Potiskum gourd bowls.
a. Karekare. 20.9 cm x 22.8 cm. UCLA MCH X83–777;
b. Ngamo. 19 cm x 20.6 cm. UCLA MCH X83–775;
c. Karekare. 22.2 cm x 24.4 cm. UCLA MCH X83–774.

Plate 21. Decorated bottle gourds:
a. Tera. 29.2 cm x 19 cm. UCLA MCH X83–690a,b;
b. Tera. 24.8 cm x 17.8 cm. UCLA MCH X83–692;
c. Tera. 14.6 cm x 9.8 cm. UCLA MCH X83–694;
d. Waja. 22.8 cm x 13.3 cm. UCLA MCH X85–53;
e. Waja. 26.7 cm x 14.6 cm. UCLA MCH X85–54.

Plate 18. Dyed gourds. Clockwise from top left: a. Settled Fulani bowl (*bodere*). 25.4 cm x 29.2 cm. UCLA MCH X83–788; b. Margi flask. 34.2 cm x 18.1 cm. UCLA MCH X83–769; c. Bata/Mboi bowl. 24.3 cm x 27 cm. UCLA MCH X83–666; d. Ga’anda spoon (*wanb’eleta*). 21.6 cm x 9.2 cm. UCLA MCH X85–22; e. Ga’anda bowl (*tɇb’sayema*). 17.8 cm x 25.1 cm. UCLA MCH X85–12; f. Settled Fulani cup (*bodere*). 12.4 cm x 11.4 cm. UCLA MCH X83–781.

Plate 19. Watercolor of gourds collected by Carl Arriens in 1911 around Kontcha, a village in Cameroon near the Nigerian border (Adamawa). Ethnologisches Bilder Archiv [#789], Frobenius Institut, Frankfurt. April 1980.

Plate 22. Gourds with a circular design focus.
a. Pidlimndi. 26 cm x 27.3 cm. UCLA MCH X83–684;
b. Dera. 19.7 cm x 20 cm. UCLA MCH X83–685;
c. Pidlimndi. 17.1 cm x 18.6 cm. Berns collection.

Plate 23. Ga'anda gourds. Clockwise from top left:
a. 14.6 cm x 21.6 cm. UCLA MCH X83–609;
b. 21.9 cm x 23.8 cm. UCLA MCH X83–613;
c. 16.5 cm x 20.6 cm. UCLA MCH X83–600;
d. 14.9 cm x 18.2 cm. UCLA MCH X83–628;
e. 17.8 cm x 25.1 cm. UCLA MCH X85–12.

Plate 24. Tera pyro-engraved gourd.
42.5 cm x 44.4 cm. UCLA MCH X83–700.

Plate 25. Bura pyro-engraved gourds. Clockwise from top:
a. 23.6 cm x 23.5 cm. UCLA MCH X83–726;
b. 14.7 cm x 14.6 cm. UCLA MCH X83–699;
c. 13 cm x 14 cm. UCLA MCH X83–689;
d. 12.1 cm x 12.7 cm. UCLA MCH X83–719.

Plate 26. Dera pyro-engraved gourd (*lib'e*).
20.9 cm x 23.5 cm. UCLA MCH X83–677.

Plate 27. Pidlimndi pyro-engraved gourd.
20.9 cm x 22.9 cm. UCLA MCH X83–676.

Plate 28. Pyro-engraved gourd bowls.
a. Chibak. 29.5 cm x 31.7 cm. UCLA MCH X83–771;
b. Kanuri. 28.7 cm x 29.5 cm. UCLA MCH X83–797.

Plate 29.
Tera pyro-engraved gourd by Jumai Pitiri Gulcoss; Wuyo.
25.4 cm x 25.7 cm. UCLA MCH X83–708.

Plate 30. Pyro-engraved gourds with a vertical design orientation. Clockwise from top left: a. Jera. 25.4 cm x 26 cm. UCLA MCH X83–703; b. Jera. 28.2 cm x 28.2 cm. UCLA MCH X83–707; c. Tera. 24.1 cm x 25.7 cm. UCLA MCH X83–701; d. Waja 24.7 cm x 25.4 cm. UCLA MCH X83–717; e. Tera. 14.9 cm x 16.2 cm. UCLA MCH X83–714; f. Waja. 23.6 cm x 24.6 cm. UCLA MCH X85–51.

Plate 31. Hona pressure-engraved gourds (*d'engyara*). a. 24.2 cm x 29.2 cm. UCLA MCH X83–643; b. 23.2 cm x 25.1 cm. UCLA MCH X83–640; c. 17.1 cm x 18.7 cm. UCLA MCH X85–41; d. 21.3 cm x 22.9 cm. UCLA MCH X83–642.

Plate 32. Karekare gourd with painted interior. Pigments. Note the alternating "helicopters" and prayer boards. 22.8 cm x 24.1 cm. UCLA MCH X83–778.

Plate 33. Pressure-engraved gourds made for the British Empire Exhibition, Wembley, 1924. a. 36.8 cm. UCLA MCH X65-8232; b. 23.5 cm. UCLA MCH X65-5233; c. 38.1 cm. UCLA MCH X65-5232. See Figs. 90–92.

Plate 35. Ga'anda pressure-engraved gourd (see Fig. 99). 17.8 cm x 22.8 cm. UCLA MCH X83-606.

Plate 34. Ga'anda brides (*perra*) dancing at the annual New Year's festival (Xombata); Ga'anda. November 19, 1980.

Plate 36. Ga'anda mud granaries (*b'ɇndɇwa*); Cijera hamlet. To the right is a man's sorghum granary and to the left is a woman's granary for storing a variety of foodstuffs. December 1980.

Plate 37. Displays of industrially-produced household items for sale in Gombe market. July 1982.

Plate 38. Mohammadu Gombe stamping designs on gourd bowls and spoons with enamel paints; Dumne. June 1981.

Map A. Nigeria

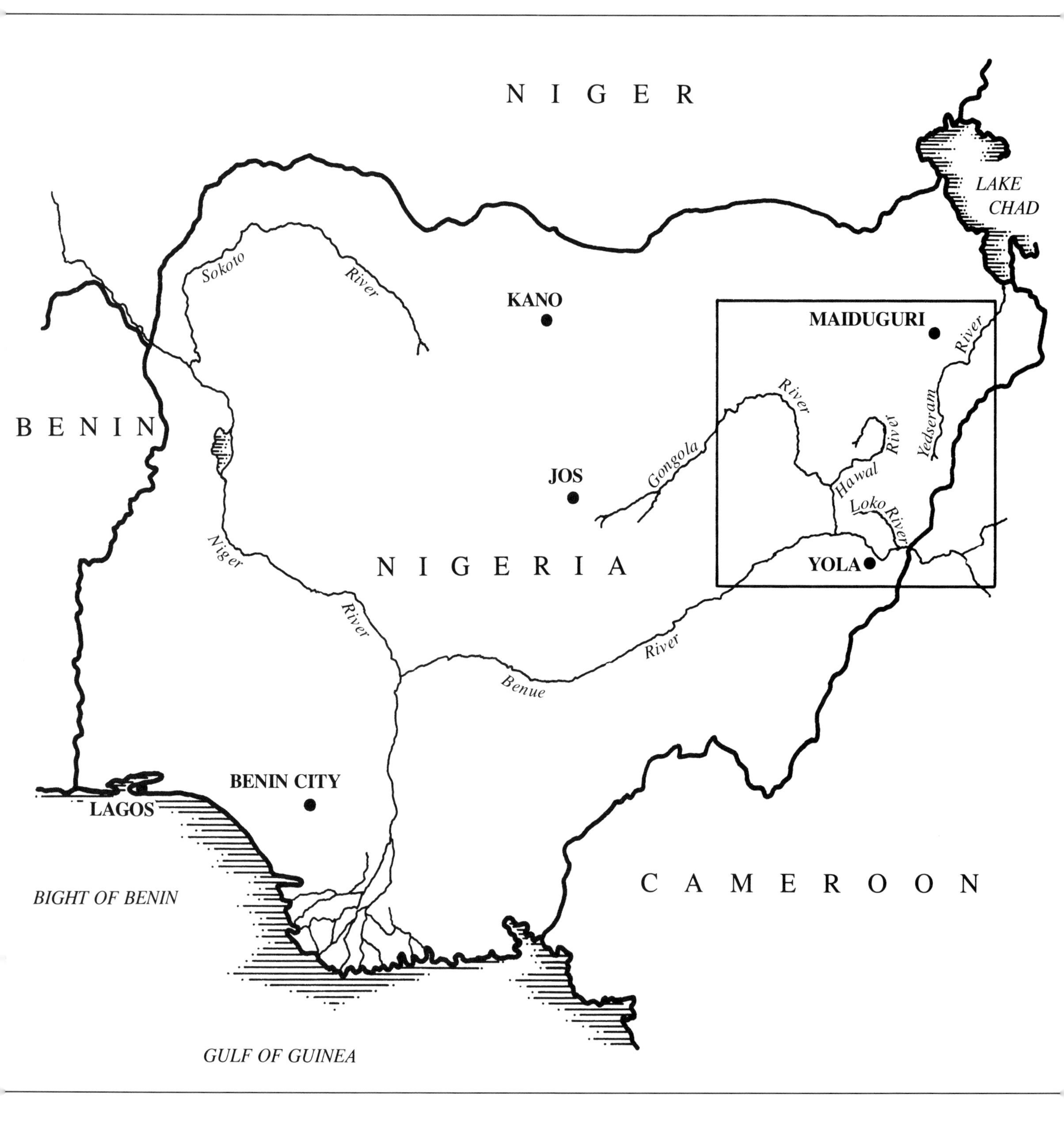

Map B. Northeastern Nigeria, with Style Areas

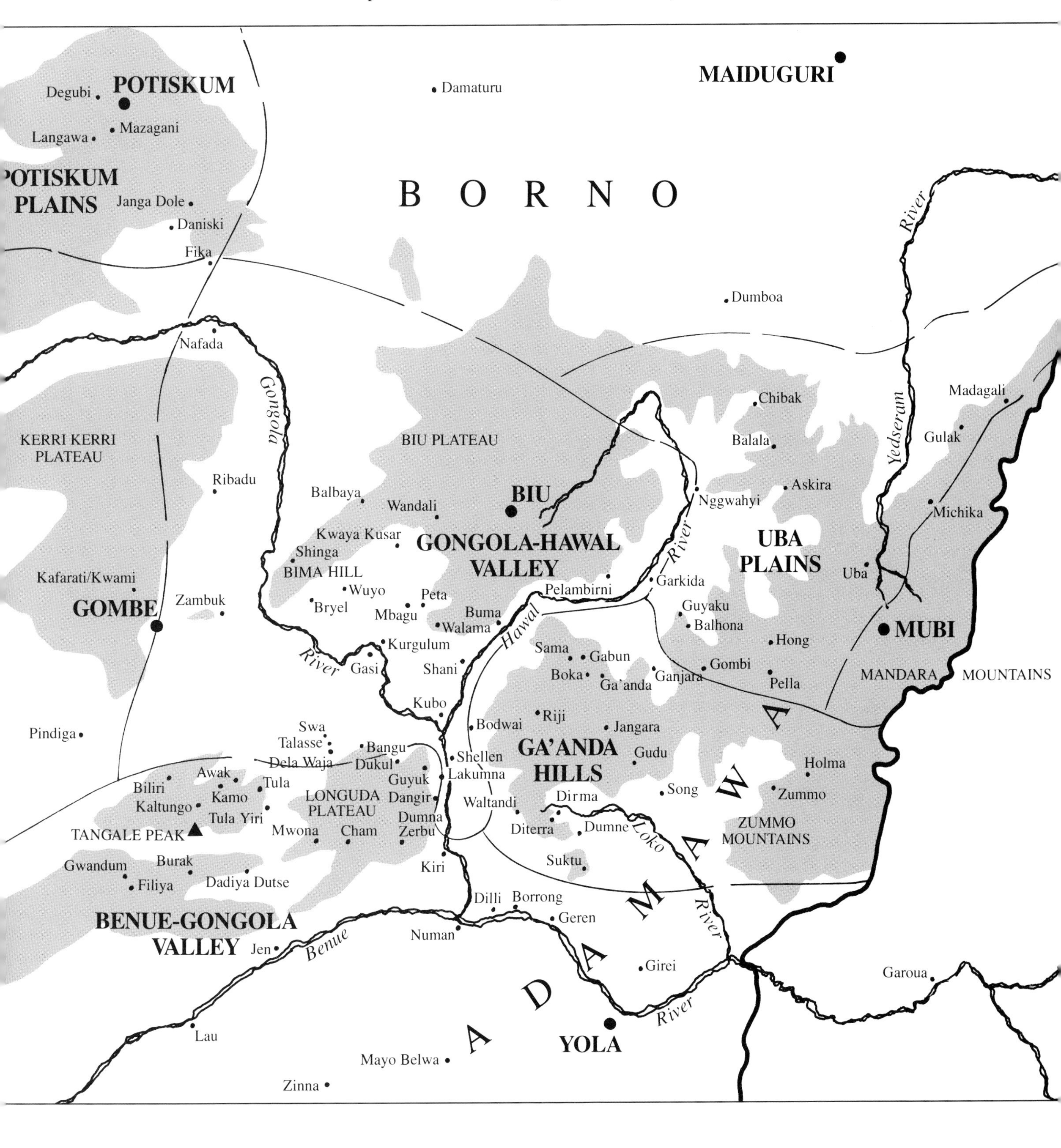

450 meters or above
(mountains, hills and strongly varied terrain)

SCALE
100 kilometers

Map C. Distribution of Decorative Techniques in Northeastern Nigeria

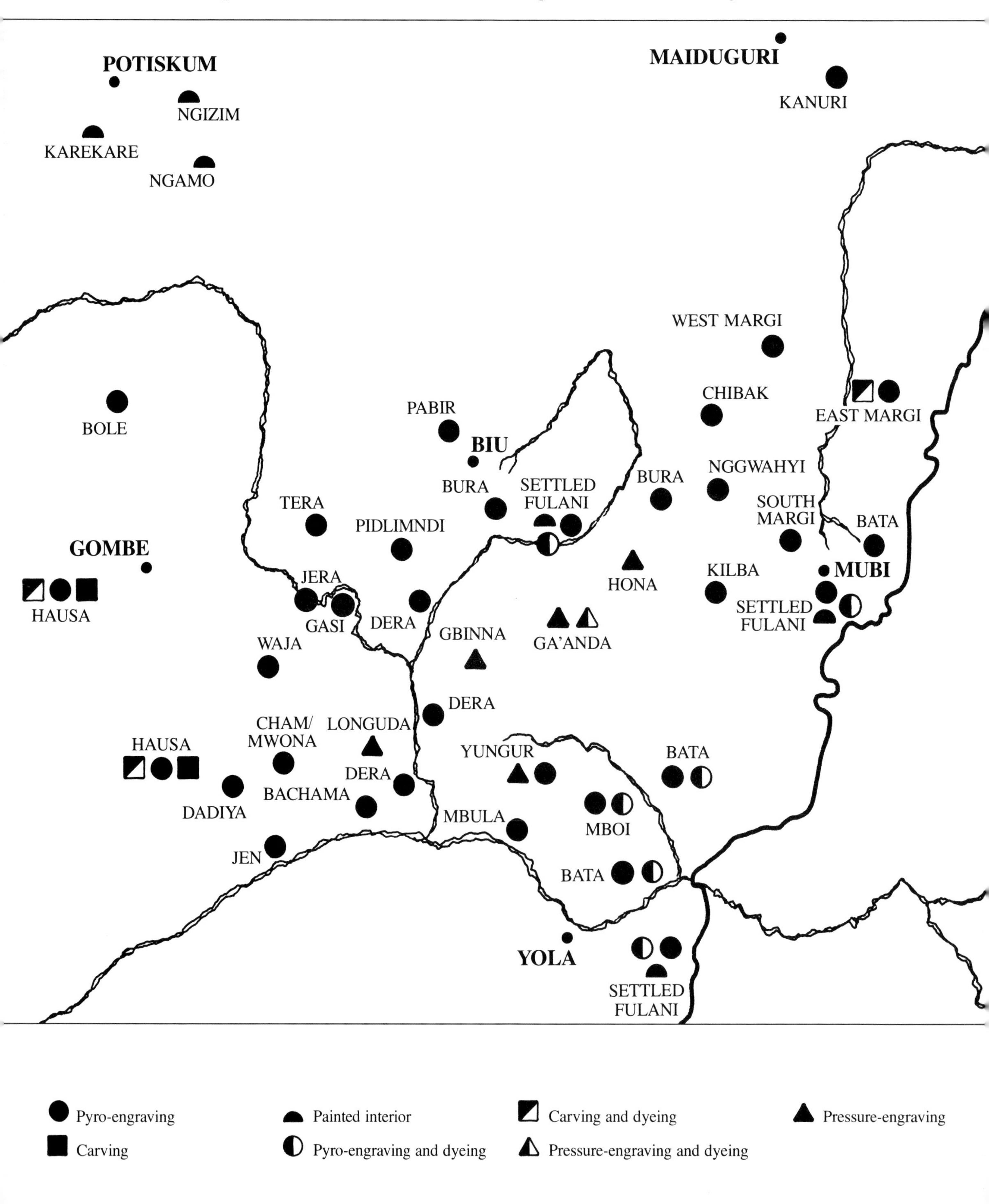

Map D. Linguistic Groups in Northeastern Nigeria

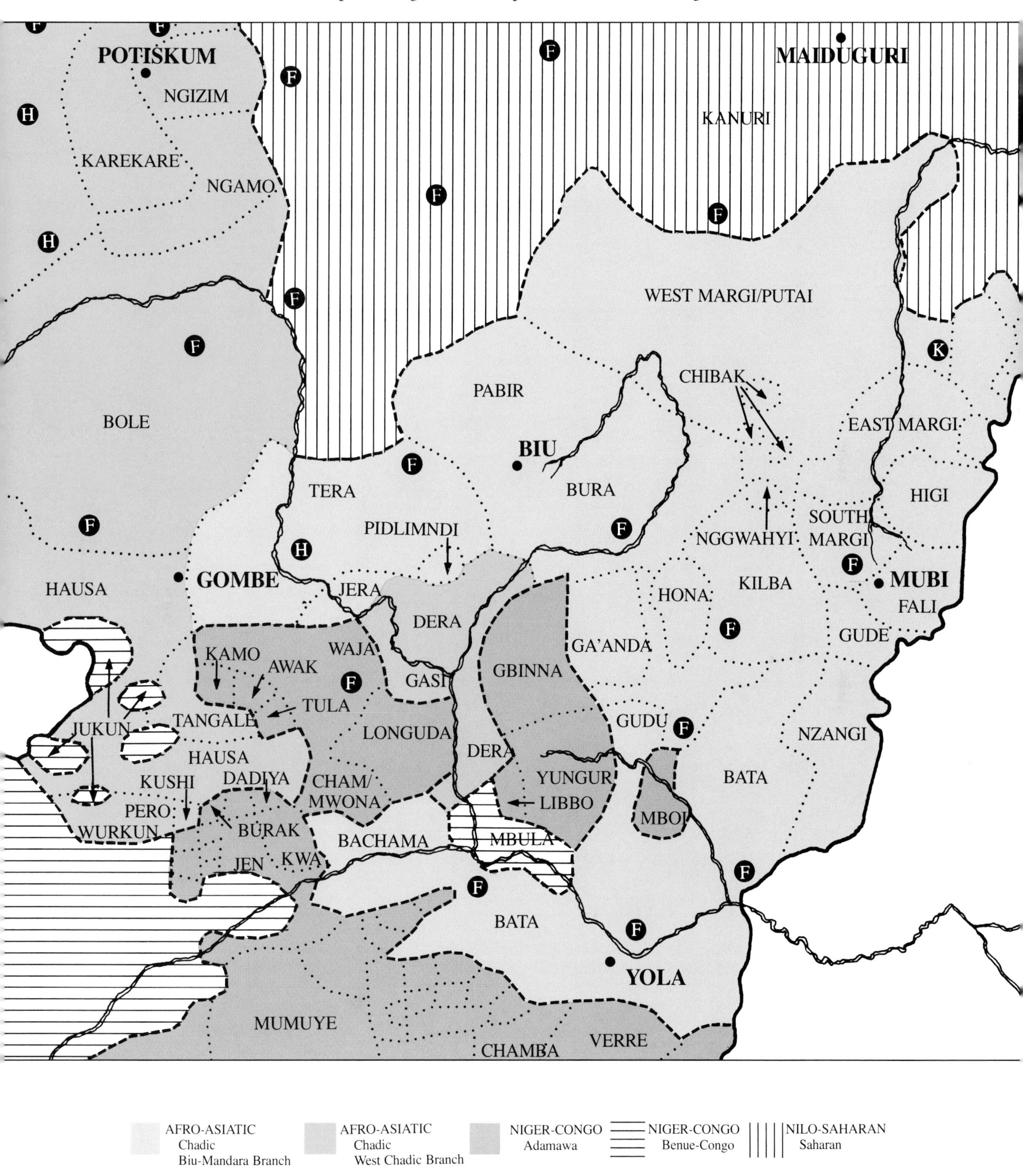

·1·

INTRODUCTION

As the fruit of one of the continent's earliest cultivated plants, the gourd has long been exploited and selectively adapted by both nomadic and sedentary peoples throughout sub-Saharan Africa.[1] The remarkable number of shapes and sizes in which it grows has made it suitable for a host of purposes, ranging from the obvious to the ingenious. The hollowed-out shells of gourds are used for storage or for serving food and drink. In combination with other materials, they become musical instruments, smoking pipes, fishing floats, or ritual regalia. The gourd's great versatility lies in its inherent properties—it is light, durable, portable, tractable, and watertight. Beyond utility, the fruit also lends itself to a rich array of decorative enhancements—from lustrous patinas, to complex patterns of incised or pyro-engraved designs, to the addition of elements as basic as fiber or as precious as beads or cowries.

Early European travelers to Africa found the gourd both novel and fascinating, resembling nothing comparable in Western cultures. Jobson, a seventeenth-century explorer, described what a gourd was and how it was used:

> Now because I speak of gourdes, which are growing things, it is fit I tell you, they doe grow, and resemble just that wee call our Pumpion, and in that manner are placed, and carried upon their walles and houses, being of all manner of different sorts; from no bigger than an egge, to those that will hold a bushell, and the necessary use they have of them, to eate, and drinke, and wash their clothes in, with divers other very fit occasions, gives the just cause to preserve them although the meate, or substance that growes within them is to bee throwne away, in regard of the extreme bitternesse, whereof the shell it selfe so savours, as no use can be made untill it be perfectly seasoned (1623).[2]

The same impulse to provide basic explanations is evident in the passages written about gourds by eighteenth- and nineteenth-century travelers.[3] Even today, with the exception of ornamental gourds seasonally available in supermarkets or calabash pipes that are popularly smoked, the remarkable shapes, sizes, uses, and modes of decorating the fruit of the gourd plant are still little-known to most Westerners.[4]

The creatively decorated gourd is ubiquitous in Africa. A number of books and articles have focused on the richness and diversity of this artistic tradition and its wide application.[5] Brief notices describing techniques and design orientations have appeared in most surveys of African crafts and decorative arts.[6] The symbolic meanings of designs have also been the subject of considerable investigation, particularly when decorative programs are representational rather than purely abstract.[7] More recent work has underscored how collections of gourds, produced primarily by and for women, can be evaluated in economic, social, and cultural terms.[8]

An examination of the literature also reveals that the most elaborate traditions of gourd decoration are confined to particular parts of the African continent. Of them, northeastern Nigeria can be singled out as an area of outstanding achievement, diversity, and inventiveness (Cover). Some peoples of this region who decorate calabash containers—such as the Fulani, Kanuri, and Hausa—are well-known.

Other groups are less familiar, yet their work has received some attention in the literature—the Ga'anda, Tera, Hona, Bata, and Yungur, among others.[9] This publication and its accompanying exhibition provide an opportunity to expand the record and to introduce a number of peoples whose creative skills were previously little-known.

The gourds from northeastern Nigeria in the Museum of Cultural History's collection come from over twenty-five different ethnic groups, and of the 275 examples made by them, no two are exactly the same. Yet, among the extensive array of techniques, designs, and compositional arrangements, each group's calabashes emerge as simultaneously distinctive and conventional, revealing as much about stylistic individuation as do African sculptural traditions of more permanent or prestigious materials. A regional survey affords the opportunity to study gourds from three complementary perspectives. First, to explore the many social, ritual, and domestic contexts in which gourds (both decorated and undecorated) are used. Second, to compare the technical and stylistic dimensions of gourd decoration in order to enhance our understanding of the evolution and role of indigenous aesthetic systems. And, last, to explore the internal factors that have conditioned the evolution of processes, motifs, and compositions and to evaluate how gourd decoration reflects patterns of ethnohistorical contact, an undertaking of special value in a region where few conventional historical sources exist.

Nearly all the gourds in this collection were decorated within the last century and many are very recent. Some gourds collected in 1970 were said to be from 50 to 90 years old; those kept as heirlooms or prized possessions were identified as the property of grandmothers (see Fig. 15a). Depending on their frequency of use and the vagaries of daily life, gourd containers can last for weeks or generations. Their decoration also endures, but this too varies with the techniques used to accomplish it.

Decorated gourds are still in high demand, despite the increasing competition from factory-made containers that have reached all but the most remote markets of northeastern Nigeria. The number of decorated gourds of recent manufacture exhibited in one small, nominally Islamic, Waja village in 1981, drawn from only a few household collections, exemplifies the strength and vitality of this tradition (Pl. 3). Even government agencies actively support its continuation, encouraging local artists to make and show examples of such secular "indigenous crafts" at annual agricultural exhibitions (Fig. 1).[10] In view of the pervasive changes affecting traditional culture in Nigeria, the persistence of gourd production and decoration is impressive.

The following study has benefited considerably from T.J.H. Chappel's systematic and thus far unparalleled ethnographic study, *Decorated Gourds in North-Eastern Nigeria* (1977). He concentrates on the work of four ethnic groups—the settled Fulani, the pastoral Fulani, the Bata, and the Yungur—who live within a fairly circumscribed sector of the former Adamawa Province. Although his research was undertaken as a part of a broad survey of material culture, the book's emphasis on gourd decoration has made it appear to be the primary and sometimes only artistic activity pursued by the groups living in this otherwise ostensibly "art-poor" region.[11] Calabash decoration, however, is only one of the many art-making activities in northeastern Nigeria. We propose that to understand fully the evolution of any one category of artistic expression, one must explore the larger universe of forms in which conventions are often established and from which innovations may be drawn. The synthetic approach favored here requires that all categories of art be considered whenever possible—which in this region include modes of pottery decoration, body scarification and ornamentation, and architectural elaboration—in order to better understand any one. It is believed that such varied information, however briefly it is presented, will provide insights into the dynamics of African art history.

BACKGROUND
Northeastern Nigeria

Geography The diversity of calabash decoration in northeastern Nigeria reflects the complexity of its geography, history, linguistics, and ethnography. The region has been divided into five main geographical "style-areas" that are correlated with the distribution of gourds in the Museum's collection: the Ga'anda Hills, the Uba Plains, the Gongola-Hawal Valley, the Benue-Gongola Valley, and the Potiskum Plains (Map B).[12]

The Ga'anda Hills, an extremely rugged upland zone, extend from east of the Zummo Mountains near Song to the escarpment bordering the east bank of the Hawal River (Pl. 1). These hills merge with the Mandara Mountains to the east, a chain of high massifs that form the frontier between Nigeria and Cameroon. Both areas are dominated by groups of inselbergs and clusters of steep-sided hills. The major peoples occupying this area are the Ga'anda, Gbinna, and Yungur. The two groups living just north and south of its borders—the Hona and Bata—

1. Display of Kilba and settled Fulani decorated gourds, Agricultural Show; Gombi. February 27, 1981.

also are included in this style-area.

The Gongola-Hawal Valley is bounded to the north by another elevated band of terrain that ends in an escarpment fringing the Biu Plateau. Within this zone of relatively high relief are a series of spectacular sandstone ridges that have been undercut by the southward-flowing Gongola River. Of these ridges, the most impressive is Bima Hill, a landmark that figures prominently in many local traditions of migration. The Dera, Tera, Pidlimndi, Bura, and Pabir dominate this confluence zone. Also included in this style-area are the Waja, who live in the sandstone hills west of the confluence.

The rugged terrain of the first two areas contrasts with the gentle, undulating Uba Plains lying to the northeast. However, they too are broken in several places by abruptly rising inselbergs, such as those near Hong, and are bordered in the east by the Mandara Mountains. The groups who live in this subregion—including the Kilba, Chibak, western Margi (Putai), and eastern Margi—are widely dispersed across the plains, extending as far south as the Ga'anda Hills and as far east as the Mandara foothills between the towns of Uba and Madagali. They mostly locate their settlements in the more broken and difficult pockets of the landscape.

A number of groups live on the lower-lying terrain of the Upper Benue River Valley. The Mbula live just east of the Benue-Gongola confluence and share the southernmost flank of the Ga'anda Hills with the Yungur. West of the confluence, a chain of steep-sided sandstone hills parallels the east/west trend of the Benue. A number of small groups live in them, including the Dadiya and the Mwona. Scattered near the towns of Kaltungo and Biliri are some of the highest hills in the area (950 meters above the plains), of which the most spectacular is Tangale Peak (Pl. 2). The Longuda Plateau is northwest of the Benue-Gongola confluence and consists of rolling plains, similar to those of the Biu Plateau. It is bounded on all sides by an escarpment that separates the Hill Longuda from the Plains Longuda.

While these four geographical subregions form a relatively contiguous belt of terrain north of the Upper Benue, the Potiskum Plains lie over 150 kilometers to the northwest. The landscape contrasts with the area lying to the south, showing very little variation in relief other than an occasional flat-topped ironstone-capped hill. A cluster of related groups—the Bole, Ngamo, Ngizim, and Karekare—are concentrated around the major town of Potiskum.

Northeastern Nigeria lies within the broad belt of savanna grassland that extends across sub-Saharan Africa. Vegetation patterns vary, however, from the wooded Guinea savanna of the south to the more open Sudan savanna shrublands of the north. Contrasts in vegetation are due essentially to differences in the mean annual rainfall, which range from 50 to 100 cm in the south to 25 to 50 cm in the north.[13] It also should be noted that the Ga'anda Hills, the immediate Gongola-Hawal confluence area, and the southern Uba Plains fall within an intermediate ecozone called the sub-Sudan that has especially complex climatic patterns and high rainfall.[14] Within each zone, arable land reflects the impact of man and his work—all communities farm during the rainy season using cyclic or "shifting hoe" horticulture. Some of the lands are also used during the dry season for pasturage (Aitchison et al. 1972:122).

Ethnography The ethnography of this region is as complex as its geography. The map shows, for example, that the highest levels of ethnic diversity generally coincide with the areas of most rugged terrain (Map B). Most of the groups living in the Ga'anda Hills, the southern Uba Plains, the Benue Valley, and the Potiskum Plains are highly decentralized and politically autonomous. All five areas are sparsely settled compared to other parts of Nigeria, and population counts rarely exceed 40–60 persons per square kilometer (Aitchison et al. 1972:text map 12). Since the colonial period many populations have moved from inaccessible hillsides to create villages and towns, although settlements are generally still dispersed. Despite the predominance of small-scale acephalous communities, a number of minor statelets emerged in the less rigorous areas where conditions were more conducive to the consolidation of authority; such states include those of the Dera in the Gongola-Hawal confluence area, the Bole further north in the middle and upper reaches of the Gongola, and the Pabir on the Biu Plateau.[15]

The four additional groups who have not yet been mentioned in this ethnogeographic classification, but whose gourds are represented in the collection, are the pastoral Fulani of the Wodabe branch,[16] the settled Fulani, the Hausa, and the Kanuri. The pastoral Fulani are a nomadic or seminomadic group whose yearly transhumant cycle has long taken them southward from Borno each dry season, passing through the Biu area on their way to the Benue. They are essentially an alien minority who have been accorded the right to graze across a wide territory occupied and farmed by a multiplicity of ethnic groups (Chappel 1977:4). Recently the contracting zones of dry season grazing land to the north have brought about a marked increase in the already substantial number of Fulani pastoralists in this region (Aitchison et al. 1972:164). The Muslim settled Fulani have dominated politically much of northeastern Nigeria since the *jihads* of the nineteenth century and the establishment of the Sokoto Caliphate. While many of the region's peoples strongly and successfully resisted their hegemony, Fulani impact has been considerable; in most areas, though a minority, they still are superimposed as the ruling elite.

The Hausa and the Kanuri represent the two largest populations living in the northern Nigerian savanna. The Hausa are concentrated in the central and western regions, but there are also Hausa settlements in the Upper and Lower Gongola valleys. There also has been a steady influx of Hausas into major urban areas of the region in recent years. As a result, in addition to the spread of the Hausa language, itinerant craftsman have brought their various trades, including gourd carving, to many of the larger market towns. The Kanuri, who represent the descendants of the original inhabitants of the Kanem-Borno empire, are generally confined to the far northeast. Recent population movements also have taken the Kanuri westward across the savanna into the Potiskum area, as well as southward to the Uba Plains flanking the Yedseram River Valley.

Languages The distribution of languages spoken in this part of Nigeria is complex and has been closely studied and mapped in considerable detail (Map D).[17] Five main language families converge in northeastern Nigeria: West Atlantic (Niger-Congo), Chadic (Afroasiatic), Adamawa (Niger-Congo), and Benue-Congo (Niger-Congo). The geographical subregions included in this study are dominated by Chadic-speaking peoples, whose languages can be further classified into two branches—West Chadic and Biu-Mandara. The divergences between them, and the groups and subgroups within them, result in languages as different as, for example, Russian and English (Newman and Davidson 1971:1). Adamawa languages exhibit comparable diversity and are spoken today by a multiplicity of groups living within the Ga'anda Hills and the complex terrain west of the Benue-Gongola confluence.

History and Art The pattern of linguistic affiliations in this region implies a similarly complex history for its peoples. In contrast to the relatively recent Fulani, Hausa, and Kanuri migrations, little is known about the prenineteenth-century history of the smaller-scale ethnic units.[18] With the exception of the open plains around Potiskum and Uba, a belt of rugged barrier terrain across the Lower Gongola Valley obstructed the expansion of major state formations such as the Borno Empire and the later Fulani emirates. Sheltered by this barrier, there emerged a constellation of largely autonomous enclaves described in the literature as "hill refuges." Because of their relative isolation and resistance to political consolidation or domination, their history and accomplishments are regarded as peripheral to the major developments of the northern Nigerian savanna. The unfortunate characterization of the peoples of this region as backward, hostile, and culturally impoverished discouraged interest in their arts, which are rarely included in regional surveys of Nigeria, both because of their relative obscurity and the tendency to focus instead on the wood sculpture and masquerades typical of the southern half of the country.[19]

The arts of the smaller ethnic groups living in northeastern Nigeria are generally treated briefly in the ethnographic literature, although a few individual traditions have been described in some detail.[20] More wide-ranging field investigations concentrating on art production were recently undertaken in the Lower Gongola-Upper Benue valleys.[21] They indicate that ceramic arts are the primary focus of sculptural elaboration in contrast to the masquerades and carved wooden figures characteristic of the Middle and Lower Benue regions.[22] Additionally, most groups living in the northeast still maintain one or more of the following secondary specializations: body scarification, gourd decoration, architectural elaboration, brass-casting, and ironsmithing. While the following discussion is primarily an overview of one remarkably versatile and vibrant artistic tradition—gourd decoration—it also seeks to bring the diverse arts and peoples of this little-known region more firmly into the purview of Nigerian art history and ethnography.

HISTORICAL NOTES

Archaeology Studying the history of plant cultivation is the work of paleobotanists who use this information to help determine stages of economic development. One systematic survey of "The History of Crops and Peoples in Northern Cameroon" has been completed by David (1976). Although focused on an adjacent area, his periodization of gourd cultivation represents the only information of its kind for the region as a whole. In a table that lists the names of crops grown by area, gourds (*Lagenaria siceraria*) are found in the

Mandaras, the Upper Benue, and Adamawa (David 1976:248). A provisional chronology dates the establishment of this crop in Northern Cameroon to at least A.D. 500. While specific information on most cultigens is still forthcoming, David concludes that the extraordinarily wide distribution of gourds within and beyond Africa "argues for their very great age" (1976:254). The earliest recorded evidence of gourd cultivation comes from excavations undertaken in an Njoro River cave, a neolithic site in Kenya dated to around 850 B.C., where one fragment of a gourd bottle decorated with dots was recovered.[23] Not only has the tradition of producing these utilitarian objects persisted over time, but so too has the impulse to decorate them.

The antiquity of gourd cultivation throughout Africa seems incontrovertible, and archaeological evidence suggests that the elaborate decoration of such containers in Nigeria dates to at least the tenth century A.D. This estimated chronology is drawn from the results of excavations undertaken at the sites of Igbo Ukwu in southeastern Nigeria.[24] Among the most remarkable objects recovered are a number of bowls cast in leaded bronze that

> appear to be based on the form of half of a large globular calabash, either cut along the maximum diameter or from the junction of the stalk and in line with it; the form of the small crescentic bowls is based on smaller, flatter, less globular calabashes cut only from the junction of the stalk (Shaw 1970 I:126).

These bronze skeuomorphs are highly elaborate, with zones of intricate, shallow relief decoration (Figs. 2,3).[25] Most also have decorative handles and spiral bosses that copy the copper handles and ornamental studs made for real calabashes (Shaw 1970 II:pls. 219–256). This suggests that the decorative tradition that served as a model for these bronze replicas is likely to have been comparably elaborate. In fact, from an area of the deposit at Igbo Isaiah, fourteen small pieces of calabash were recovered, each decorated with straight and curvilinear incised lines (Shaw 1970 I:250). While not suggesting a historical relationship, the bronze bowls in the Igbo Ukwu corpus bear interesting parallels to gourds carved recently by peoples living a long distance away (such as the Longuda in the Benue Valley; Fig. 4; cf. Shaw 1970 II:pls. 220–223). They show that the subtractive methods of gourd decoration (burning or engraving lines into the surface) may have been translated, through the prodigious skills of the brasscaster, into fine filligreelike overlays. This possible decorative inversion, which represents considerably more effort than directly impressing lines into a wax model, correlates with the style of other Igbo Ukwu castings. It also suggests that by means of a precious material, gourd bowls were functionally elevated to the status of prestige objects.

Additional archaeological evidence exists in the form of a small bronze libation cup excavated south of Lake Chad, just east of the Nigerian border, at the site of Midigué. It represents a unique find within the so-called "Sao" corpus, and because of the irregularity of its curved rim, is described as skeuomorphic of a small gourd bowl (J.-P. and A.M.D. Lebeuf 1977:69; fig. 91). The replication of this common object in bronze also elevates it to the status of regalia associated with the "sovereignty and power" of the Sao.[26] Unlike the Igbo Ukwu bronzes, however, this bowl lacks an equivalent decorative overlay and has only a simple rim frame and bifurcated handle. While there is some question as to the chronology proposed for Sao occupation of the Chad Basin, the Lebeufs' reconstruction suggests that this bowl was probably cast after the thirteenth century A.D. (1977:194).[27]

The beginning of gourd decoration most certainly predates the chronologies proposed for Igbo Ukwu or Sao, but when this art emerged in the northeast is not known. What seems likely, however, is that the ornamental potential of the gourd's surface was not fully exploited until after a shift from a stone-based to an iron-based economy, which in the Upper Benue region would have occurred after the fifth or sixth century A.D. (David 1976:237). The advent of iron tools and the invention of pyro-engraving not only offered more technical possibilities to the artist, but also would have provided an efficient means for cutting the gourd into useful shapes as well as depithing it.

Linguistics Linguistic evidence confirms the antiquity of gourd cultivation. The Fulani language, for example, includes at least twenty different terms that refer specifically to the size of a gourd, its use, or both.[28] In the Hausa language, more than twenty-three terms for gourds and pumpkins are popularly used. Most of the smaller ethnic groups represented in this study also have multiple names for gourds, many of which distinguish their shape or their use. What is most distinctive about the vocabularies associated with gourds is that botanical varieties and functional applications tend to be differentiated by unique word stems or morphemes. The fact that these words are not cog-

2. Large bowl cast by the lost-wax method in imitation of a hemispherical decorated gourd. Igbo Ukwu. Bronze. Diameter of rim: 26.4 cm. National Museum, Lagos [54.4.22].

3. Small bowl cast by the lost-wax method in imitation of a halved elliptical decorated gourd. Igbo Ukwu. Bronze. Length: 17.8 cm. National Museum, Lagos [39.1.5].

4. Longuda gourd bowl (*kwarawa*). 22.2 cm x 22.5 cm. UCLA MCH X83–658.

natic—i.e., they are not derived from a common original word—suggests that the selection for diverse shapes of gourds was not recent and that it was culturally imperative to differentiate clearly between them (pc: R. Schuh, December 1984).[29] This linguistic evidence is especially striking in light of the fact that only two words—"gourd" and "calabash"—are used by Western peoples (and they are used interchangeably) to identify a wide range of selectively bred fruit shapes and sizes.[30]

Not surprisingly, gourds figure prominently in cosmological as well as metaphorical thought. The Hausa identify the universe as a spherical gourd, the two halves of which represent the sky and the earth, with the rims where they join as the horizon (Parrinder 1967:42).[31] The Fulani compare a gourd filled with smaller gourds to "the sky filled with stars," or even "Allah and the stars" (Chappel 1977:22). The round shape of a gourd container may indeed suggest the "completeness" of the universe; yet, it is more likely to be its ubiquitousness and importance to traditional culture and economy that translates it into a significant metaphor.

References to gourds are common in proverbs and riddles. For the Hausa, spicing one's speech with aphorisms, parables, and idioms, classified together as "proverbs" (*karin magana*), is a popular way of projecting mastery of language (Skinner 1968:79). The pithy expressions that follow give some idea of the essential importance of gourds to the Hausa:[32]

1. *Duma yana rad'an mad'achi*
   "The bitter gourd jibes at the mahogany tree,"
   i.e., the pot calls the kettle black.
2. *A rarrabe da d'an duma da d'an kabewa*
   "One will distinguish between the bitter gourd and the sweet,"
   i.e., we shall be able to distinguish between true men and false.
3. *Inda aka iske duma a nan a kan fafe shi*
   "Where the gourd is found there it is scooped out,"
   i.e., wherever you meet him he'll be ready to oblige you.
4. *Mod'a ba iri ba ce*
   "A gourd cup isn't seed,"
   i.e., there is no harm in appointing a person to an official position even though it is not hereditary to him.

5. *K'waryar aro kowa ya fasa ki ya dunka ki*
   "A loaned calabash, whoever breaks you must mend you,"
   i.e., a thing I treasure is like the apple of my eye.
6. *Gora in bai yi tsiran kowa ba ya yi tsiran igiya*
   "The calabash if it saves no one, will save the rope,"
   i.e., . . . because the latter is tied onto it.
7. *Gyartai ya ci sarauta ya ce ban da tuna baya*
   "The mender of calabashes became a great one and said, 'Let us not remember the past,'"[33]
   i.e., he wished to forget his humble beginnings.
8. *Al'amarin nan k'warya rufe ke nan*
   "This matter is covered by a gourd,"
   i.e., this matter is a mystery.
9. *Shi ma zai bi mu, cikin k'warya d'aya za mu ci abinci*
   "He will follow us; we eat together out of one gourd,"
   i.e., he will be our close ally.
10. *Ruwa ake kamar da bakin k'warya*
   "Its raining like the mouth of a gourd,"
   i.e., it's pouring cats and dogs.
11. *Kowace k'warya tana da murfinsa*
   "Every gourd has its lid [matching half],"
   i.e., everyone has the chance to decide what suits him and what does not.
12. *K'warya ta bi k'warya, in ta bi akushi, sai ta fashe*
   "A gourd follows a gourd; if it follows a wooden food bowl, then it will break,"
   i.e., don't tackle what is beyond you.

·2·

AN ETHNOGRAPHY OF GOURD USE

Gourds are usually cultivated on farms during the rainy season along with major food crops, or they may be planted directly within compounds where they are encouraged to trail over fences and the thatched roofs of houses (Fig. 5). The gourd plant, a climbing annual with very rapid growth, ripens between four to six months after planting. It requires heavy rainfall during early stages of growth and high temperatures and substantial sunshine during the period preceding harvest (Hodge 1982:15). Among some groups, farming responsibilities are divided and both sexes harvest gourds; among others, such as the Ga'anda, men have exclusive control over their cultivation and preparation. In some villages and towns, cultivation is the work of certain households that serve the needs of the immediate community. In larger towns, some households provide modest quantities of gourds to sell in local markets during the late dry and early rainy seasons (Fig. 6). In major centers, large numbers of calabashes are sold throughout most of the year (Pl. 4).

Gourds are grown in four basic shapes in northeastern Nigeria: globular, flattened-globular, tubular, and bottle-shaped. Within these categories, gourds of many different sizes and contours were developed by means of selection, and their degree of diversification is a credit to the skills of the African cultivator. Each breeds true to type, due essentially to structural variations, especially in the fibrovascular bundles (Chappel 1977:8). For example, the Hausa cultivate at least four spherical gourds of varying diameters—the smallest ones are made into ink wells (*kurtun tawada*; Fig. 7), while the largest ones (*gora*) are fitted with handles and used as floats, such as those that figure prominently in annual fishing festivals (Pl. 5).[1] Other shapes reflect similar adaptations, particularly those of bottle gourds whose profiles and stem alignments vary considerably (Fig. 8).[2] The simplest type grows in the form of an elongated cylinder and gradually widens at the base until it takes on the profile of a straight, narrow-necked flask. Some varieties grow with a long constricted neck and a sharply flaring mouth; others look more like short "dumbbells." The stems are also bred to curve or to grow to extreme lengths.

Usually picked when they reach full maturity, gourds may be harvested when unripe if an intermediate size or shape is desired. Because the hard shell of the fruit is all that is required, the spongy, fibrous contents are first removed either by leaving the gourd in the sun until the pulp dries and shrivels, or by soaking the gourd in water until the pith rots. The gourd is then cut open, depithed, and scraped clean (Fig. 9). How the gourd is opened determines the shape of the resulting container and the ways it can be used (Fig. 10). Globular gourds are usually cut through the stem-axis to create two equal hemispherical bowls (Fig. 10a). A deeper container is produced by cutting the gourd unequally (Fig. 10b). Flattened spherical gourds are generally also cut through the stem-axis, producing two elliptically shaped bowls well-suited for drinking purposes (Fig. 10c). The same gourds bisected latitudinally are used to make wide, rather shallow bowls with lids; the short, projecting stem umbilicus forms a useful handle on the lid (Fig. 10d). A bottle-shaped gourd can be bisected along its stem-axis to create two ladles or spoons (Fig. 10e). The long, narrow stem

5. Bottle gourds growing in a field behind a woman carrying a gourd container; Uganda. 1959.

6. Prepared gourds for sale at the weekly market in Dumne (Yungur). June 1981.

7. Hausa ink well (*kurtun tawada*). 9.5 cm x 10.2 cm.
UCLA MCH X83–793.

8. Varieties of bottle gourds;
drawn after examples in the UCLA MCH collection.

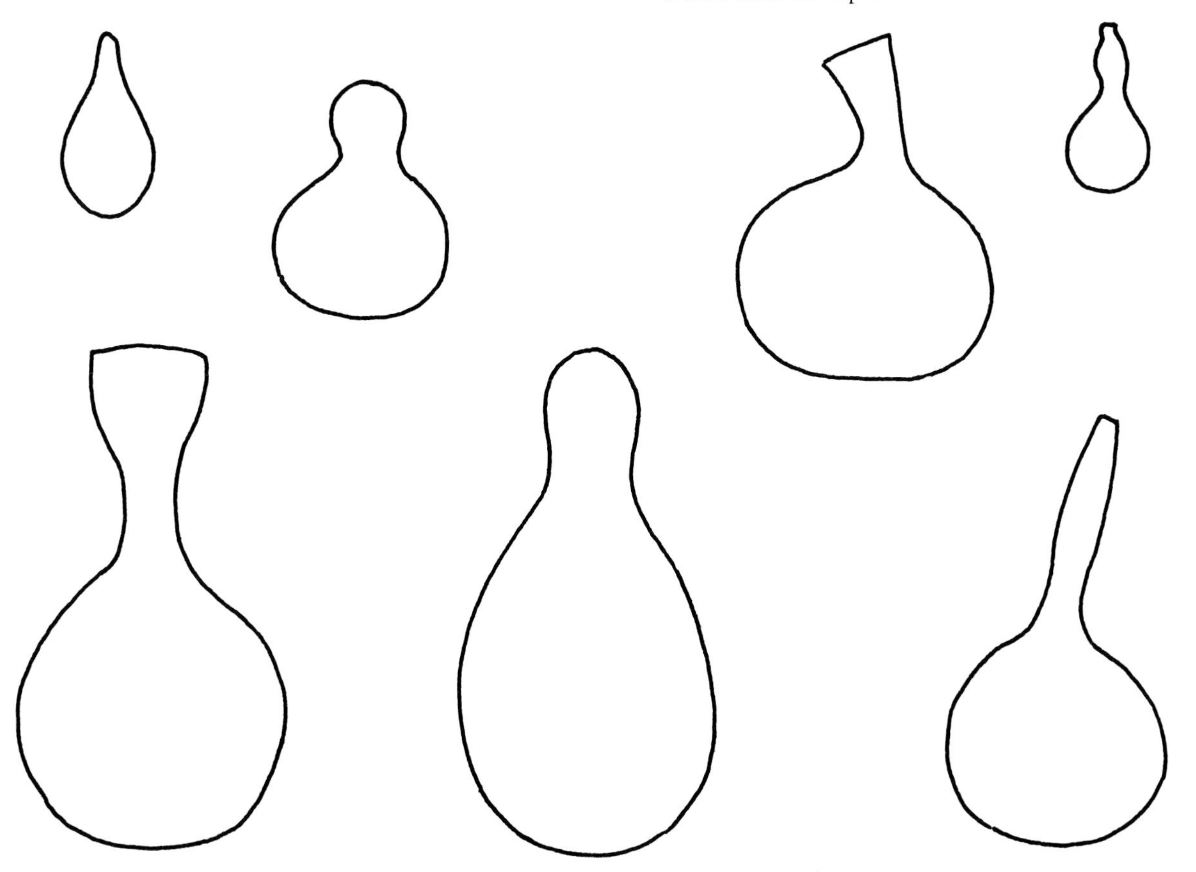

9. A man cutting, depithing, and scraping his harvest of gourds in the field where they were grown; between towns of Gombe and Numan. 1970.

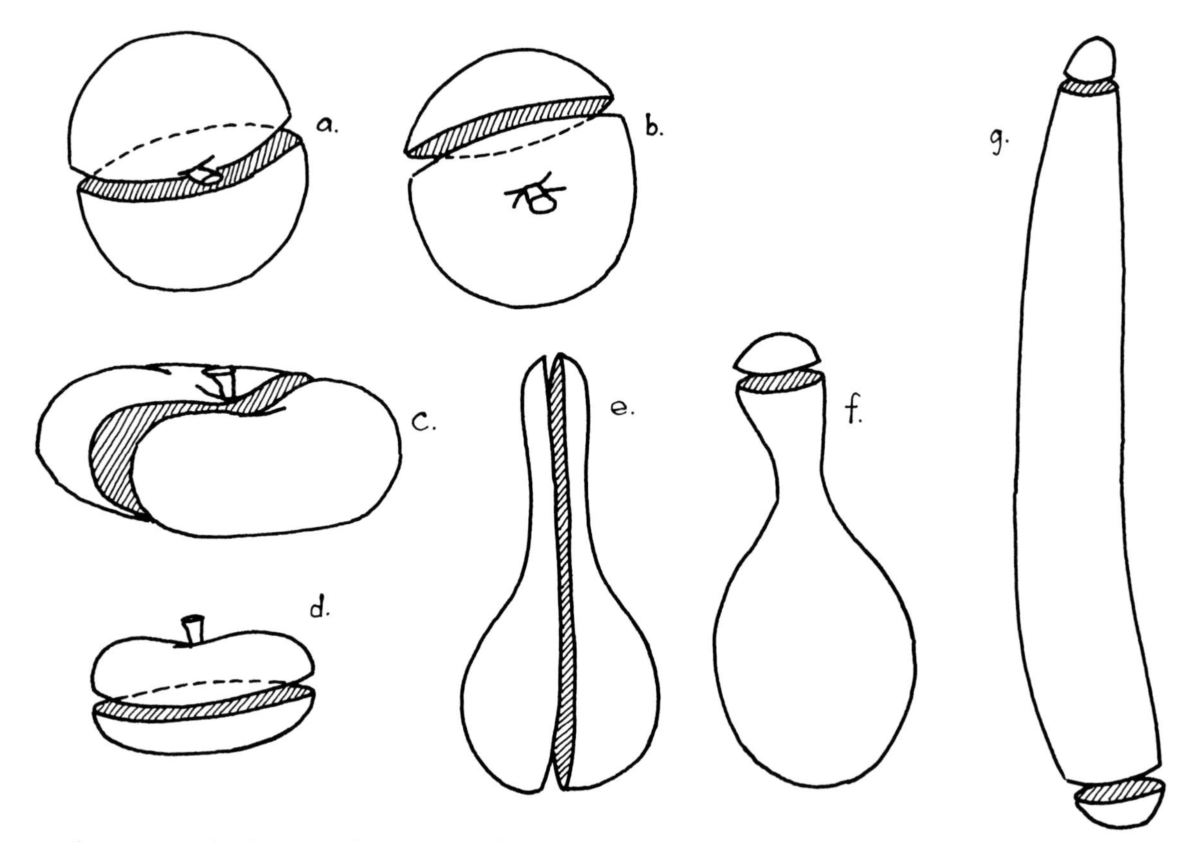

10. The way gourds of various shapes are cut for cleaning. The resulting shapes determine their context of use.

can be used as either a handle or a pouring channel.

All gourd varieties are easily scraped clean once an axial cut is made.[3] However, the procedure for depithing a bottle gourd and some tubular varieties that are not bisected, is more difficult and often quite time-consuming. The stem end is generally sliced off so that small pebbles can be dropped in and shaken around to loosen the pulp and seeds (Fig. 10f).[4] What cannot be freed in this fashion may be removed painstakingly with a thin prod. Once cleaned, the open-mouthed bottle gourd serves as a convenient flask for holding liquids or other materials. Also, an aperture can be cut into the body leaving the stem end as a handle. Or, if the opening is cut larger, the stem serves as a handle for a deep ladle. Tubular gourds, which often grow to great lengths, are cut at one or both ends depending on how they are to be used (Fig. 10g).

Once the desired shape of any gourd is achieved, it is dried until the outer shell has thoroughly hardened. This may take from one to two months depending on the size and maturity of the fruit when harvested (Hodge 1982:15); the drying period cannot be shortened by placing a gourd directly in the sun, as this causes the shell to become brittle and crack (Chappel 1977:8). When properly hardened, a gourd is ready for decoration or for use.

Gourd decoration is primarily a woman's part-time dry season occupation, although this work may be done whenever there is spare time. Any woman with the necessary skills may decorate gourds, but relatives and neighbors of well-known artists usually prefer to commission the work from them. The prepared gourd is usually supplied by the patron, and the effort expended in decoration was traditionally rewarded with food, usually guinea corn, with the gourd in question serving as the measure. More recently, cash has been used as payment, but the cost still is largely determined by the size of the gourd.[5] For example, a Bole man working in 1971 was charging three pence ($.04 U.S.) for decorating a small gourd, one shilling ($.14 U.S.) for a medium-sized one, and one shilling six pence ($.21 U.S.) for a large one.[6] In the Margi town of Balala, male and female artists working on commission were paid five shillings ($.65 U.S.) to decorate a large gourd and four ($.56 U.S.) for a medium-sized one.[7] By 1980, women were asking between one and two naira ($1.80–$3.60 U.S.) for a "custom" decorated gourd; those made for sale in larger markets were somewhat less expensive. While it is evident that some artists today do earn a certain amount of cash selling decorated gourds, among some groups, artists still prefer to accept commissions that are paid in foodstuffs. Even women who own large gourd collections rarely allude to the cost of their production because most were originally received as gifts (Chappel 1977:24).

DOMESTIC CONTEXT

Gourd containers are essential items of household equipment. Hemispherical gourd bowls in a variety of sizes are dominant household utensils, and decorated containers of this type are represented in the Museum's collection in the greatest number. While most women in northeastern Nigeria prefer decorated calabashes, plain unembellished gourds are kept for certain tasks, such as holding ingredients during cooking or ladling water from storage pots. Ornamented bowls ranging from ten to thirty cm in diameter are primarily used as receptacles for serving food and drink. Because of their insulating properties, gourds with naturally thick walls are especially useful for holding or covering hot foods. Women select containers of particular diameters for serving specific foods—the smallest cups are used by mothers to feed their infants; those that are medium-sized hold soups or stews; and those that are large best accommodate the glutenous mass of porridge made of sorghum (guinea corn), rice, or millet that forms the staple of most diets. Each is usually named according to its size and specialized function.[8] For example, a Dera woman serves her husband meals in a carefully arranged vertical stack with a small soup basin (*lib'e d'ila* or *lib'e d'oyara*) at the bottom, followed by a hemispherical bowl containing porridge (*lib'e muni*), and covered at the top with another bowl that serves as a lid (*lib'e tib'ekokumat*) of equal or slightly larger size (Fig. 11). This arrangement helps keep the food warm and makes for an attractive presentation. Water is served in a separate, hemispherical container (*lib'e b'uta*) that often has been rubbed inside with red ochre and oil to seal the surface. Among the Dera and other groups, the interior surfaces of calabashes with thick walls are decorated so their designs are visible while a person is eating (see Fig. 51). Decorated gourds directly enrich the experience in which they are employed—i.e., even eating or drinking is better if the container used is decorated.

Calabash bowls of large diameter (40–60 cm) are widely found as carrier-containers. They are used to transport produce within the compound, from the fields to the compound, or from the compound to the local market, balanced on top of the head by means of a fiber- or cloth-bound ring or carried against the back of the shoulder, depending on cultural norms (Frontispiece). Pastoral Fulani women use deep

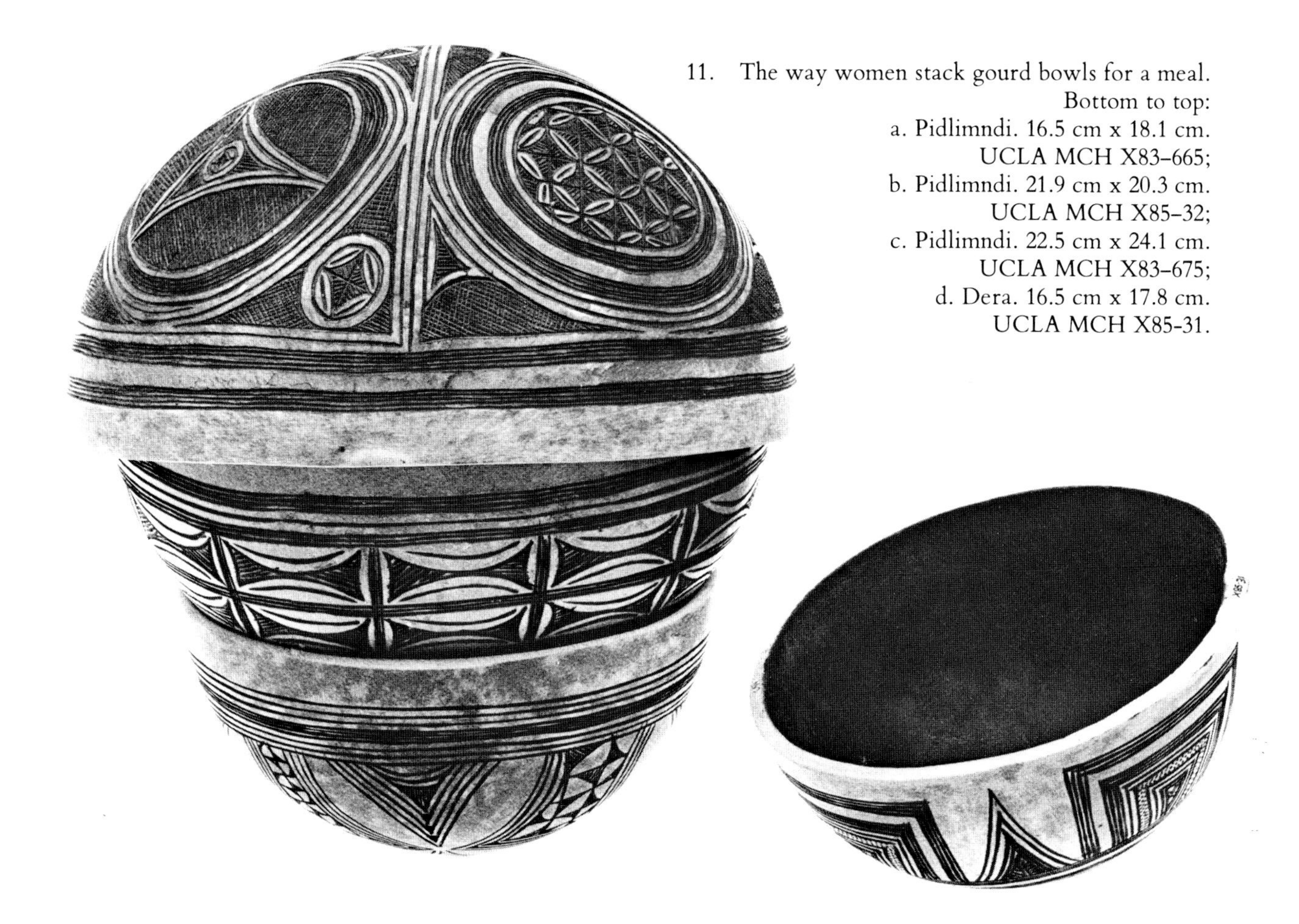

11. The way women stack gourd bowls for a meal. Bottom to top:
a. Pidlimndi. 16.5 cm x 18.1 cm. UCLA MCH X83–665;
b. Pidlimndi. 21.9 cm x 20.3 cm. UCLA MCH X85–32;
c. Pidlimndi. 22.5 cm x 24.1 cm. UCLA MCH X83–675;
d. Dera. 16.5 cm x 17.8 cm. UCLA MCH X85-31.

bowls made from unequally cut gourds for safely transporting quantities of milk over long distances to markets (Fig. 12). Large hemispherical containers are also useful as portable storage bins; their shape allows women to stack them as "drawers" or suspend them from ceiling rafters in rope nets.

In addition, hemispherical bowls of various sizes, often left unadorned, are employed as measures for the exchange of foodstuffs. The Ga'anda, for example, differentiate four gourd sizes for the purpose of bartering with grain. One size, called *teb'wanketa* ("calabash-hoe"), holds the amount of sorghum considered equivalent to the value of an iron hoe blade. A second bowl, called a *teb'fenda* ("calabash-grindstone"), is used to measure out the amount of grain to be ground for a single day's porridge. The last two gourds, *mafatib'a* and *kapatib'a*, refer to the largest bowls.

The Hausa, who call any cut circular gourd *k'warya*, have long used these containers to establish certain widely accepted trading standards.[9] A calabash of needles equals 100 needles, a calabash of kola nuts equals 100 kola nuts, a calabash of garlic equals 100 bulbs, and so forth. *K'warya* is commonly used as a metaphor for the diameter of a round room—e.g., "the calabash of this room is equal to. . . ."

While bowls cut from globular gourds predominate in daily use, the three other varieties of gourds—bottle-shaped, flattened and tubular—also have a number of specialized applications. Calabash spoons cut from bottle gourds make handy ladles for serving soup or utensils for drinking the sweet thin porridge that is often shared out of one large serving bowl (Fig. 13). Bottle-shaped containers are used not only as flasks for liquids, but also to hold various medicines and cosmetics (Pl. 21; Fig. 14). Smaller ones are filled with pebbles and used as babies' or dance rattles. They are also used for holding seed, so that during sowing the contents can be shaken into the hand before being placed in the ground (Wente-Lukas 1977a:47). Tiny "bottles" (10 cm) that grow with a short narrow neck are used for administering enemas to infants.[10] Large varieties are sometimes used by pastoral people as butter churns.[11] And, quite the opposite from these seemingly mundane applications, bottle-shaped gourds with extremely attenuated stems are used in the Cameroon Grasslands as elegant palm wine containers.[12]

Elliptical bowls are highly prized by certain northern Nigerian groups for drinking beer because of their convenient shape (Fig. 15). Ga'anda women reserve a large elliptical bowl ("gourd of the household," *tɇb'kɇnnda*) for serving their husbands a special breakfast of heated beer mixed with ground sesame or melon seed. In the south, very small palm wine cups also have been documented, and a bronze skeuomorph recorded by Nicklin in the Cross River area is so flat and narrow that it can be easily cupped in the hand while drinking (1982:49; fig. 1). It also is likely that the bronze crescentic bowls excavated in Igbo Uwku were once used as ritual palm wine vessels whose prototypes were bisected elliptical calabashes (cf. Fig. 3).

The last type of gourd, grown in a long tubular shape, is mostly used by Moslem women for covering their hands after a dye paste made from henna leaves is applied (Figs. 16,17).[13] The gourd protects the hand until the dye saturates the skin and serves as an ornament during the process. Short cylindrical gourds also can be used for serving soup or drinking liquids.

While they are used primarily by women, men also own gourds of which the bottle variety appears to be the most common. Decorated or plain, they are carried as canteens during a hunt or a day's farming. They can be attached by means of a leather thong to a belt or to a quiver (sometimes made of a long tubular gourd) slung over the shoulder (Fig. 18). A small, hemispherical bowl can serve as a "dipper." Moslem men carry gourd flasks for religious ablutions. Although not seen in this part of Nigeria, smoking pipes made from bottle-shaped gourds are popular across the continent, particularly in the eastern and central regions. Bottle gourds bred to grow in a range of shapes and sizes, some in striking configurations, are especially suited for making water pipes.[14] Men also use small round gourds fitted with lids as snuff bottles.

12. Pastoral Fulani milk vendor carrying a decorated gourd bowl; Girei district. 1965.

The preceeding inventory shows that gourds are highly versatile containers. Another desirable feature of gourds is that they are handily mended. Most groups have devised special techniques for joining the breaks that usually occur at the rim edge (Fig. 19). Gourds must be repaired quickly, because they can warp unevenly making later corrections more difficult. Ga'anda women, for example, use two methods for repairing a cracked gourd—one called *mɇkta*, using vegetal fibers, and the other called *timda*, a more recent technique utilizing spun cotton thread as well as European yarn or nylon. In both techniques, a series of holes is bored or burned with an awl on either side of the break; the fibers are then pushed through the holes with either a porcupine quill or a needle and the broken edges are drawn together. The neat row of woven sutures achieved by careful lacing often translates the repair into an attractive adornment. In fact, one bottle gourd collected among the Mwona was decorated with large embroidered letters using a modified "mending" technique (Fig. 20). The four letters—"PAWA"—were stitched with cotton thread across an unbroken surface, a difficult procedure that must have required a considerable amount of skill. The word itself—a phonetic transcription of the English pronunciation of "power"—may allude to national or ethnic pride, an interpretation reinforced by the way the letters have been so masterfully "written."

Even if a calabash breaks irretrievably, pieces of it are exploited for various ingenious uses. Larger fragments can be used as containers or as small trays. Pieces of gourd may be shaped and their edges rounded to be used as scoopers for ladling thick porridge from the cooking pot to the serving bowl. Among a number of groups, large curved fragments are effectively employed as turntables in pottery-making; crescent-shaped pieces also work well for scraping and smoothing the walls of vessels. Furthermore, the Ga'anda, for one, use the triangular point or straight edge of a small gourd chip to impress designs in the clay surface, finding them a good device for achieving tight, crisp decorative patterns (see Fig. 118). Beyond these contexts, Malzy lists a number of other interesting ways she saw gourd fragments being exploited: as scoopers for sweeping, as canoe bailers, as components of field scarecrows, as covers for snuffboxes, as seals for rattles, and as "scratchers" for circumcised men who are forbidden to use their hands (1957:12).

13. Decorated gourd spoons. Left to right:
a. Dera. 37.6 cm. UCLA MCH X83–732;
b. Dera. 19 cm. UCLA MCH X83–674;
c. Hona. 18.1 cm. UCLA MCH X83–638;
d. Hona. 22.5 cm. UCLA MCH X83–633;
e. Bura. 16.5 cm. UCLA MCH X83–693.

14. Decorated bottle gourds:
a. Waja. 22.8 cm x 13.3 cm. UCLA MCH X85–53;
b. Tera. 29.2 cm x 19 cm. UCLA MCH X83–690a,b;
c. Waja. 26.7 cm x 14.6 cm. UCLA MCH X85–54.

15. Ga'anda beer bowls (*tɇb'kɇnnda*).
a. 20.3 cm x 32.1 cm.
UCLA MCH X83–603.
This gourd was decorated by the grandmother of a highly reputed Ga'anda artist (Witebar), and was said to be over 90 years old;
b. 18.4 cm x 27.5 cm.
Berns collection;
c. 18.7 cm x 30.2 cm.
UCLA MCH X85–8.

16. Henna gourd. Bura. 40 cm.
UCLA MCH X83–761.

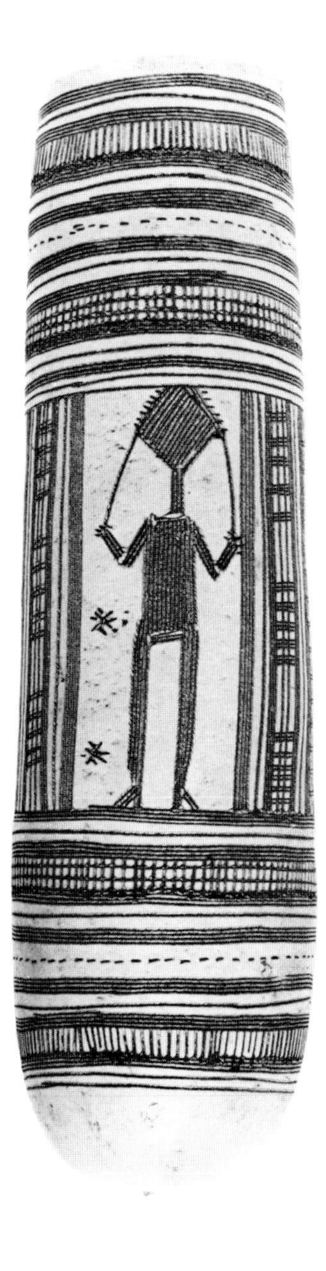

17. Henna gourd (*janturu*). Pastoral Fulani.
Russell and Maxine Schuh collection.

18. Waja man dressed as a warrior for a "staged" performance of Kwolodangeri, a festival generally held to celebrate the death of an elder; Dela Waja. Note decorated gourd dipper suspended from the warrior's quiver. February 1982.

19. Tangale woman mending a large gourd bowl using twisted fiber rope; Biliri. April 1982.

20. Mwona bottle gourd with letters "P-A-W-A" stitched across body. Cotton thread and indigo dye. 29.2 cm x 18.1 cm. UCLA MCH X83–792.

SOCIAL AND SACRED CONTEXTS

This compendium of domestic use gives some idea of the gourd's fundamental value. The gourd's domestic utility enhances its ability to communicate social as well as economic values. Among most of the groups represented here, large collections of decorated gourds are an essential part of bridewealth payments and dowries. Among the Karekare, for example, traditional marriage customs dictate that a bride take to her new husband as many as thirty large water containers, forty medium-sized food bowls, and thirty smaller drinking vessels, all decorated. The considerable effort or expense involved in assembling such a collection indicates the importance of a woman's transition to full marital status. Collections are often amassed over a number of years, drawing on the services of many close relatives. While forms of payment may vary, it is likely that an artist's investment of time is repaid through reciprocal social obligation. It is interesting to note that among the Yungur, who require the groom (or groom's family) to give the bride gifts of decorated gourds, men have recently taken up the art.[15] Because the products are needed by men, it is apparently legitimate for men to make them even though Yungur women traditionally provided the service for their male relatives.

The culmination of marriage agreements often involves more than simply an exchange of gourd capital; some form of community display is also required that usually includes all of the household items that have been amassed. Such public exhibitions demonstrate the woman's ability to set up an independent household and her husband's capacity to provide. The size of the collection varies, however, as does the kind of ceremony with which it is displayed, both at the time of a girl's wedding and thereafter. Among the Ga'anda, a special wickerwork basket (*can'lan'nda*) is made to present all the decorated gourds collected by the bride (Pl. 7). Their neighbors, the Yungur, hold a final marriage ceremony (*kahn sektesa*) where the bride dances with a similar basket (*gilango*) balanced on her head (Chappel 1977:14). It is filled with all the calabashes and other gifts given to the bride by her suitor and relatives. While wearing a special string skirt (*gabo*) made of twisted baobab pith dyed with red ochre, the Yungur bride is presented with a final gift by her husband—usually a decorated gourd—thus signaling that he may now make sexual advances.[16] This gift, called a *dinge gabo* ("gourd of the skirt"), was explained by Chappel's informants as "the gourd which loosens the bride's skirt." If the bride accepts it, it signifies that she will abandon her girlhood tradition of wearing string aprons and adopt a covering of cotton strips that marks a woman's elevation to full marital status. Berns' Yungur informants added that before the bride formally joins her husband, she lines up a series of decorated gourd bowls (*dinge gabo*) extending from his compound entry to the door of their conjugal room. Still wearing her string skirt, she walks along this path, straddling the gourds with her feet. This act similarly acknowledges her formal acceptance of marital responsibilities.

The importance of a woman's gourd collection in establishing social and economic status is nowhere clearer than among the pastoral Fulani.[17] A young girl sets up an independent household and gains full marital status two or two and a half years after the birth of her first child. Her formal entry into her husband's homestead is marked by two complementary ceremonies, both called *bangtal*, that publicly establish the respective economic positions of each partner. At the first, a man is given a herd that is his symbol of economic independence. At the second, a girl is elevated to the status of wife and mother through the formal gifting and exhibition of household items including gourds, mats, cloth, and a bed. The large number of milk gourds in her dowry marks a woman's right to milk her husband's herds. After this inaugural public display, a Fulani woman always keeps a part of her "trousseau" carefully wrapped up for ceremonial occasions in a pack called a *kaakel*. A *kaakel* can hold fifty (or more) elaborated calabashes nested in sets of ten—each set completely encased in a woven mesh—and all supported on two wooden poles (Beckwith and van Offelen 1983:20). This portable bundle is kept near her bed and is ready to be loaded onto a pack ox whenever the group moves. The "working" portion of her calabash collection remains accessible for daily use. Whenever a new camp is made, the wife arranges her working gourds on a special platform "in order of decreasing size and running in the feminine direction north to south" (Chappel (1977:10). This procedure allows a woman to project publicly the dynamics of her changing economic position—the size of her gourd collection increases with each addition to the size of her household. Also, whenever a girl is born, more decorated calabashes are added to the *kaakel* pack in preparation for the time a woman will pass them along to her own daughters when they marry. It is clear that among the pastoral Fulani, a woman's prestige is intimately tied to her reproductive powers, i.e., the increase of the size of the domestic work force. The size of both gourd collections thus reflects the economic viability of the household and a woman's satisfaction of essential familial and social obligations. The elaborate displays of the contents of *kaakel* packs at communal celebrations held each wet season provide an additional arena in which women can publicly proclaim their successes (Pl. 8). Such dramatic statements about personal wealth complement the way a man's presentation of cattle declares his economic position.

Gourd containers are ideally suited to a nomadic lifestyle because they are light, durable, and portable. And because they can be tightly nested, it is possible for women to accumulate them in enormous numbers. Chappel has convincingly demonstrated that the *kaakel* pack and its elaborate display are objects of considerable aesthetic attention (1977:24). The amount of care and skill lavished in arranging the *kaakel* represents more than a desire to produce "a convenient and manageable load." The fact that its contents are carefully laid out for public exhibition shows it is to be appreciated as more than a reflection of its owner's economic and social prestige. Whether embellishing a domestic scene or a ceremonial occasion, these gourds are treasured for their aesthetic value.

There is ample evidence that the ornamentation and display of decorated calabashes serve the same purposes for other groups living throughout the northeast. The aesthetic and expressive importance of gourd display is elaborately resolved among the Tera. One year after a woman is married, a low mud platform (*kankame*) is built along the inner wall of her sleeping room. The platform provides a permanent stage for the exhibition of calabashes, pottery, and other household items (Pl. 9).[18] Pots can be stacked up to four high and two deep, with each pile surmounted by hemispherical gourd bowls nested in groups of three or four. This kind of arrangement does not make for easy accessibility. The Tera room illustrated in Plate 9 reflects the conscious aestheticism of these exhibitions despite the fact that it contains no mud platform. It reveals especially well the careful placement of each item, including the circular mats (*faifai*) on which gourd ladles are aligned in pairs. The possibility exists that the permanent, linear arrangements of Tera household items may have been influenced by the seasonal exhibitions of the pastoral Fulani. Although the Tera tradition is waning today, some women still enhance the interiors of their rooms by painting their mud platforms with red and white pigments and laying tesselated potsherd pavements dotted with encrustations of cowries.

These Fulani and Tera displays make a dramatic statement about a woman's marital status. At the same time, individual decorated gourds are ubiquitous emblems of women's household roles, obligations, and accomplishments. They can embellish any of the domestic contexts in which they appear; the rather stark interiors of rooms are enriched by calabashes drying on a rack, covering water pots, or even hanging on a wall. Among the groups living in the Potiskum area, large gourds are fitted with twisted leather thongs both to make them easier to handle and to allow them to be suspended so that the interior painted surfaces show. This ornamental aspect explains why so much attention is lavished on decorating a surface that most likely is not visible when the container is being used. In fact, the painted interiors of many gourds from this area are dull and dark from the smoke of cooking fires in the rooms where they are displayed. Margi women suspend tall stacks of calabashes bound up in twisted rope nets (*talagu*, "dish road") in their rooms (Fig. 21). Other groups, such as the Dera, store their collections in large wickerwork baskets. Mbula women mount clusters of gourds on tall pottery stands that are positioned around the interior of their rooms (Fig. 22). While storage may be the ostensible purpose in most cases, the Mbula example shows that gourds are intentionally arranged so as to be pleasing when looked at and thus worthy of admiration. Moreover, as these objects also represent original dowry or bridewealth payments, they may be regarded as indicators of status in much the same way as they are for the Fulani.

While most of the contexts just described represent private aspects of display, there are times when decorated gourds are publicly visible and likewise carry a socioeconomic and aesthetic message. Despite the widespread adoption of industrially manufactured deep enamel basins, women from most groups often prefer to use large gourd containers to transport food and other provisions (Frontispiece; Fig. 12). As Plate 10 and Figure 23 show, women also use hemispherical bowls as "bonnets" for protecting their infants from the hot sun or from the rain.[19] Many groups use gourds in this way and, among the Bura, a baby's "layette" typically includes a number of "hats" in various sizes that are often passed around within a family (Fig. 24; B. Rubin 1970:23). Yet, as the illustration here reveals, ornamented gourds also make a bold visual statement. Today they form a part of an aesthetic display that involves

21. Margi resist-dyed gourds (*kabaki*) stacked inside a rope binding (*talagu*); Gulak. 1971.

various machine-made wrappers and baby slings, as well as imported blouses, polyester headscarves, and plastic jewelry. An early photograph of a Pabir woman included in Meek shows that before the adoption of factory-made goods, a baby was carried in a leather satchel covered with a cloth binding (1931 I:143). The mother wore only a large homespun wrapper so that the ornamented gourd bonnet was a striking addition to other traditional aspects of self-display, such as facial scarification, elaborate coiffure, and beaded or metal hair, lip, and ear ornaments. Such ensembles (either traditional or modern), of which the distinctive designs on a gourd helmet or carrier-container form an integral part, function as ethnic identification. A woman's membership in a particular group and her marital status are conveyed in explicitly visual terms that have special relevance in areas that are ethnically complex and fluid. Nomadic Fulani women lavish considerable attention on their appearance before traveling to markets—sometimes fifteen or more kilometers away—to sell their wares.[20] The milk gourd on a woman's head is carefully integrated into a program of personal adornment intentionally designed to be visually pleasing, and some of the complex motifs engraved on the gourd are often depictions of facial scarifications (Fig. 25).[21] There is some evidence that particular gourd designs once served as modes of identification for individual Fulani lineage groups; by repeating certain facial markings, group membership could be visually reinforced (Chappel 1977:30).

Visual and symbolic associations of women with gourd containers extend to festival contexts. Many groups in this region organize elaborate communal gatherings to celebrate life-cycle transitions. As indicated above, a new bride's collection of gourds is frequently exhibited once she marries. Additionally, young women who participate in marriage festivities often carry a decorated gourd in one hand as a part of the costume that distinguishes the occasion. Among the Dera, a series of special dances is performed each year at Ilela, a festival traditionally held to honor newly initiated boys, and girls who had completed an elaborate program of body scarification. Initiation and scarification are no longer practiced in Dera communities as essential prerequisites to marriage. Ilela, however, has been preserved as a social event where boys and girls who will soon marry, publicly display their finest ornaments and exhibit their dancing skills. During the festivities, a line dance called Bwalin is performed by young brides, each of whom vigorously jingles a decorated calabash bowl (*lib'e wa*) with leather straps threaded through brass rings (Pl. 13).

22. Mbula pottery stand (*nkwarin'nda*) with decorated gourds (*gilu kwar*) as they would be displayed inside a woman's room; Dilli. 1982.

23. Bura woman using a gourd bonnet (*dambelam*) to protect her baby from the sun; Wandali. 1969.

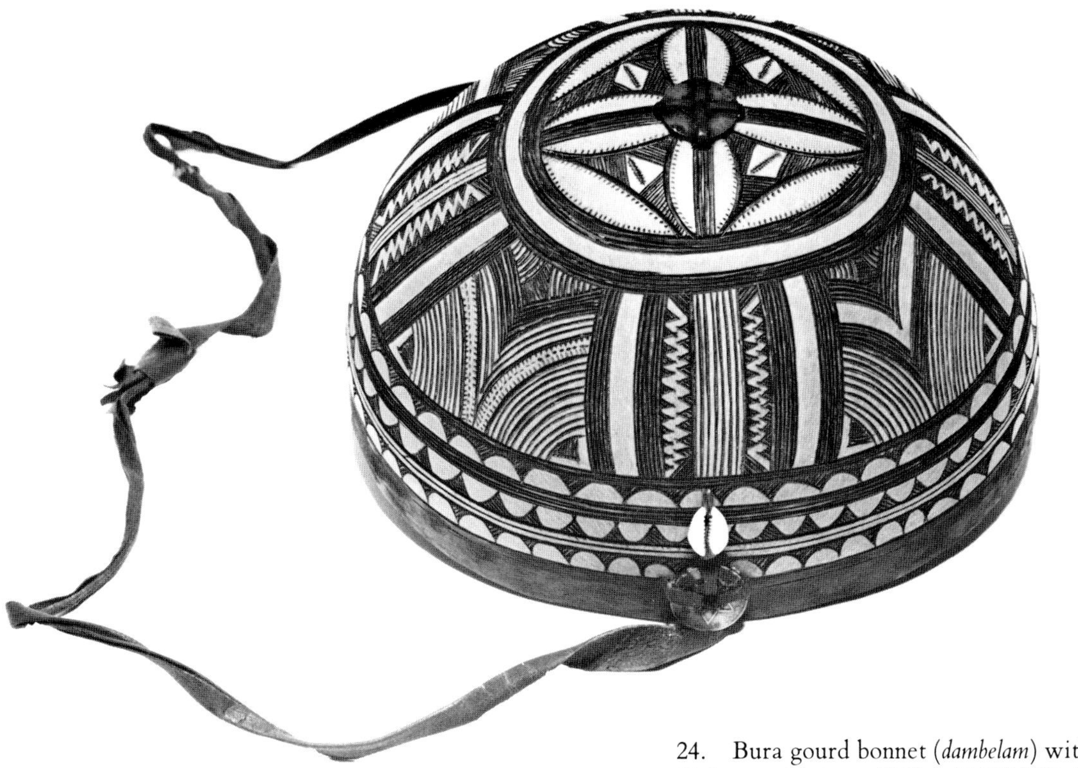

24. Bura gourd bonnet (*dambelam*) with leather straps. 22.5 cm x 24.3 cm. UCLA MCH X83–760.

25. Pastoral Fulani milk gourd (*tuppande*). 33.3 cm x 36.2 cm. UCLA MCH X83–785.

Gourds are worn by male participants at a preliminary dance held before Xono, the Yungur initiation ordeal. Chappel regards the gourd helmet, "covered with a layer of beeswax in which ears of corn are embedded, and also the red seeds of the Crab's eye vine," as essentially an armature for the additive elements that carry either a decorative or symbolic meaning (1977:14).[22] The corn may be a reminder of a young man's obligation to maintain an independent household by farming, as the gourd bowl reminds a young woman of her domestic responsibilities. Chappel notes that at the completion of Xono, the father gives his son a hoe and his mother gives him a gourd (1977:14). Again, the message conveyed concerns the fulfillment of familial obligations. The fact that the boy usually passes the gourd given by his mother to the girl who has brought him food during his period of seclusion (whom he may not necessarily marry) suggests that the calabash bowl is associated with female nurturing.

Although gourd bowls that figure in dance contexts are frequently associated with women and women's roles, an interesting exception is the calabash dance rattle (*kichibyok*) carried by Dadiya men at Kal festivals held every five years to celebrate the coming of age of young men (Fig. 26). The male aspect of this calabash is established by the iron stirrup-shaped handle attached to the gourd with strips of leather. The crest of the handle takes a distinctive and elaborate form, with a wrapped extension and an open basket of twisted strands threaded through small rings.[23] This iron-handled dance attribute complements other iron regalia carried during Kal and other Dadiya dance festivals celebrating the activities of men. For example, forged iron daggers (*nyansanye*) are used to acknowledge publicly the skill and bravery of young men during battle or in the hunt (cf. Fig. 27).[24] The handle of a ceremonial dance sickle (*jen'nyi*; Fig. 28) carried by men during Kal, even more closely resembles the iron portion of the *kichibyok*. The association of iron regalia with the role of men as protectors and providers has transformed the gourd "dance rattle" into an emblem of male rather than female status.

Hemispherical gourds are held during annual agricultural rites, again usually by women. For example, Ga'anda women carry small gourds during Xombata, the major event when thanks are offered for the passage of the year and for a successful harvest. At a similar festival performed by the Dera called Menwara or Menjoli ("beer of acclamation"), women from the chief's household dance with decorated gourds or with elaborate pottery vessels balanced on their heads (Fig. 29). In Shani, a

26. Dadiya dance rattle (*kichibyok*) carried by men during Kal festivals. Iron, leather, cotton thread. Height of gourd and handle: 40 cm. UCLA MCH X83–796.

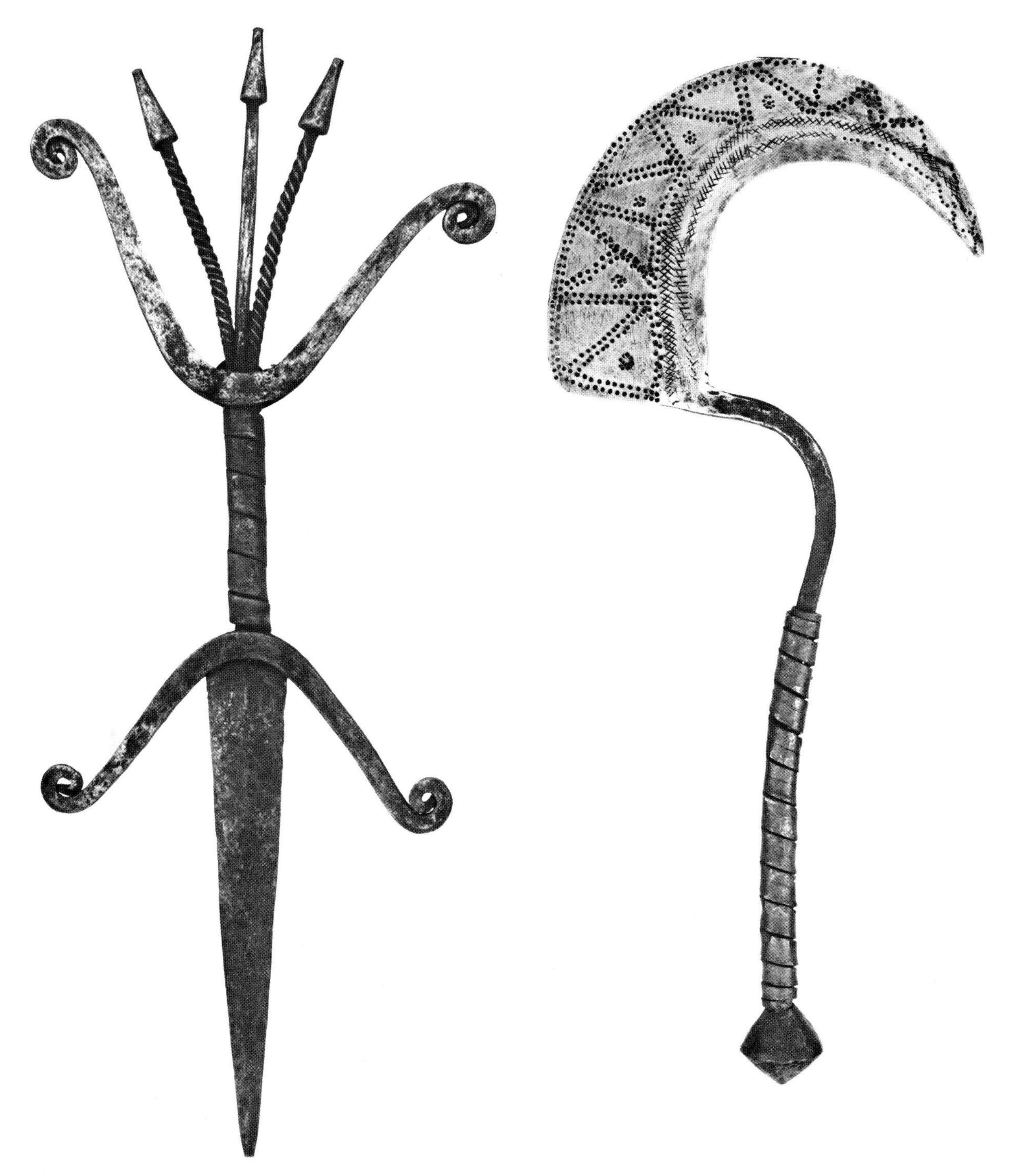

27. Tula ceremonial dagger (*tochile*) carried by young men in dance contexts, identical in form and function to the Dadiya *nyansanye*. Iron. 38 cm. UCLA MCH LX83–48. Promised gift of Jeanne and Jim Pieper.

28. Dadiya ceremonial dance axe (*jɇn'nyi*) carried by young men in dance contexts. Iron. 39.1 cm. UCLA MCH LX83–47. Promised gift of Jeanne and Jim Pieper.

29. Dera women from the chief's household dancing with ornamental pots (*wunda*) and calabashes (*lib'e*) during Menwara; Shani. December 10, 1980.

major Dera town, this event has largely been translated into a Moslem festival honoring the chief. Nevertheless, certain traditional elements have been retained to distinguish the "ethnicity" of the proceedings. In addition to the procession of women bearing the pots and calabashes that signify their role in the economic success of the previous year, a program of dance performances from surrounding localities is staged, including Bwalin (see Pl. 13).

Infusions of Hausa/Fulani culture, as well as Western modes of dress and adornment, have strongly influenced the appearance of both men and women in festival contexts. Early photographs however, especially those taken in more remote and hilly regions, show how prominently gourds were once used as expressive ornaments on ceremonial occasions. One illustration of the Fali, who live in the Bossoum region just across the Nigerian border in northern Cameroon, shows a funeral rite at which every woman and girl is carrying a gourd in one hand and a digging stick in the other—all that is "worn" other than highly conventionalized forms of jewelry and body decoration.[25] As ornaments, these essential tools may be emblematic of women's roles in an agricultural society—the stick may signify their participation in cultivation while the gourd marks their responsiblity in converting produce into food and beer. By carrying these insignia, women symbolically offer spirits of the deceased the same nourishment they provide during life (Huet et al. 1978:106).

The significance of women as food preparers should not be underestimated and the gourd container, used to present the endproduct of this effort, is a meaningful emblem of this role. Men depend on women for their meals, as well as for the beer that households must contribute to various social and ceremonial activities. According to the rules of etiquette followed by most groups, only women can cook and prepare the staple foods of the diet.[26] Additionally, women often are the sole cultivators and gatherers of supplementary foodstuffs used in preparing the soups and stews that accompany the main carbohydrate staple. These labor-intensive activities are the primary means by which women maintain considerable household leverage and authority. Men who are, or who become, bachelors must find a female to cook their daily meals, putting them at a distinct social disadvantage. A husband will often go to great lengths to secure his wife's loyalty and to reduce her temptation to seek a more satisfactory arrangement. Meek's observations of the Gabun (Gabin), a Ga'anda subgroup, provide one particularly revealing example of the relative strength of a woman's household position:

> If a wife chooses to leave her husband she can usually secure another without difficulty, and I observed among the Gabin many elderly men who were wifeless and were dependent for their food on the wife of a younger brother. Many elderly men have to do their own cooking. Wives are the beer-makers, and they do not hesitate to prevent their husband distributing beer to friends of whom she does not approve. A wife may even prevent her husband from attending a cooperative day's work on a neighbor's farm (at which beer is freely distributed) if she considers that he would be better employed at home. A case came to my notice in which a man's wife prevented him from carrying out an order of the chief, until he had finished his work of hoeing her farm. I also came across an instance of a wife living with her husband and children in the home of her parents, because she could not endure her husband's relatives (1931 II:380–381).

Thus a woman's gourds are not symbols of domestic drudgery, but rather reflect the pride she can take in her social and economic position. Indeed, when a Longuda woman decides to run off with another man, she secretly sends all her best calabashes to her new home before absconding in the dark of the night (Meek 1931 II:338).

The Yungur maintain an interesting custom that reinforces the conceptual link between gourds, women, and household stability. At second funerals of elderly men or women (Wora) held annually during the early planting season, wives of the deceased's sons gather together and ceremonially break one decorated gourd each. They then retrieve a large fragment and dance with it during the Wora festivities. The significance of this gesture is twofold: on the one hand, it marks the separation of the dead from the realm of the living; on the other, it conveys a message to the deceased about each wife's commitment to look after and feed his or her offspring.

Communal beer drinking, ranging from the purely social to the highly sacred, is another context where gourd containers figure prominently. In most areas, beer made from guinea corn or millet is served in a calabash bowl, often sealed on the inside with red or black pigment. Like palm wine in southern Nigeria, beer is an "important social lubricant, indispensable for hospitality and conviviality" (Northern 1984:134). Beer made from grain has also long served as an important nutritional supplement. It appears that beer consumption in traditional times was restricted to a number of sanctioned social or ceremonial occa-

30. Ga'anda lineage heads listening to praise-singing before drinking ritual beer brewed to honor a deceased community leader during Kwefa; Gabun. November 24, 1980.

sions. Substantial quantities of beer were available during the farming cycle, when households provided it in payment for assistance given during days of communal clearing, planting, and harvesting (when caloric boosts were most necessary). More recently, however, brewing beer has become a profit-making enterprise, and women sell it to neighbors or take it to local markets. The hub of men's social interaction, whether in the village or in the market, is usually wherever beer is being sold (Pl. 12). Although round bowls are often used, elliptically shaped gourds are especially suitable for drinking purposes because the narrower end helps prevent spillage.

While decorated gourds are usually used in secular contexts where beer is consumed, there is a strong prohibition against using them during ritual activities or for sacrificial libations when sharing beer is a primary means of cementing ties with powerful spirit forces. This may be because the exclusion of women from sacred contexts is extended to the objects made by their hands and with which they are so intimately associated. In fact, among the Ga'anda, the same proscription is extended to the food bowls used regularly by ritual priests—they may never eat from the decorated containers a Ga'anda wife typically uses for serving her husband's meals.

Ceremonial beer drinking is a key part of the Kwefa festival held by the Ga'anda in the dry season following the death of a chief or a high-ranking community leader.[27] In Figure 30, lineage heads representing the main families living in the Kancikta ward of Gabun (one of the three Ga'anda subsections) are preparing to honor the spirit of a recently deceased community leader. Each elder is given one undecorated elliptical gourd (*sambata*) filled with a special kind of sweet, unfermented guinea corn beer (*yemnda*) brewed only on such ritual occasions. This beer is also poured into the ceramic vessels (*hlefənda*) where the spirit of the deceased has been localized. They are removed from the room where they are kept after the burial of the corpse to participate in the event held in their honor (Fig. 31). What is significant about this procedure with reference to the subject at hand, is that a small calabash bowl is always kept over the mouth of the pot. While it functions to prevent dangerous forces from entering

31. Ga'anda flutists serenading the vessels (*hlefənda*) containing the spirit of the deceased during Kwefa; Gabun. Note the gourd bowls protecting the mouth of the vessel and serving as the spirit's drinking "cup." November 24, 1980.

the *hlefenda* unbidden, it also suggests that, like their living counterparts, the spirits of the deceased must be provided with an appropriate container for drinking beer. To further distinguish their ancestral identity, the gourds kept on *hlefenda* are not entirely plain, but are dyed red, a color that has strong ritual connotations for the Ga'anda and other groups. It is associated with various activating substances used for invoking the aid of spirit forces. Red hematite (Fe_2O_3) is one such substance and it may be a substitute for animal blood that is used for sacrificial propitiation.

The use of gourds in contexts where powerful forces are invoked is fairly widespread. The suitability of the gourd for such situations may be due to its essential and acknowledged role in household processes. David and Hennig's observation about the contextual versatility of pottery containers also applies to calabash containers: they would hardly have figured in ritual or sacred contexts unless they were already highly valued in domestic and social spheres (1972:28). The fact that most belief structures are organized to perpetuate processes of material survival suggests that objects critical to the latter can assist either actually or symbolically in the efficacy of the former. It should be noted, however, that gourds are rarely used as containers for directly localizing spirit beings, a function in this general region reserved primarily for ceramic vessels of various kinds. For example, the Yungur use terracotta vessels as permanent respositories for spirits of ancestors worshiped during annual ceremonies (see Fig. 125). During preburial rites, however, a gourd coated with red ochre is placed on the head of the corpse where it remains until the interment a few days later (Chappel 1977:16). This gourd is carefully preserved so that if a member of the household falls ill and it is determined that a particular ancestor is responsible, the gourd can be used as a drinking cup by the patient. It is believed that the positive forces associated with the deceased are captured in the gourd and it is thereby effective as an instrument for curing the patient. It is interesting that while the same custom is described by Meek, he claims that gourds smeared with red mahogany oil were only placed over the heads of old women who had died (1931 II:464). In this case, it can be argued that the gourd is an appropriate receptacle for concentrating the forces associated with women and their role in sustaining household well-being. Both Meek and Chappel have proposed, however, that the gourd is used as a substitute for the ancestral skull that might once have been employed for this purpose.[28]

Another spirit-charged context in which gourds play a significant role was documented among the Mbula. They use large bottle gourds (*du*) as emblems of membership in the powerful men's secret society, Ngala.[29] Young men who are candidates must complete two to three months of intensive training before they qualify for entry. Once initiated, Ngala members are able to cure any illness, physical or otherwise, that afflicts an individual or an entire community. Gourd bottles, *dungala* ("gourds of Ngala"), are also used by members as dance rattles during festivities held by the society.

In every Mbula community, the hamlet head, his spirit priest, his earth priest, and his war chief each maintain a public display area (*tankul*) for keeping ritual paraphernalia. The *tankul* illustrated in Figure 32 belongs to the earth priest, Menzali, who oversees the forces responsible for crop fertility in the Dilli area. The focus of this shrine is a large fig tree, in front of which a three-pronged branch has been erected for suspending certain ritual items. The prominently displayed *dungala* signify the Menzali's membership in Ngala.

If an Mbula leader dies, all of his personal belongings are displayed on his grave—dug directly in his compound—for one year. The example in Plate 11 was constructed to honor the highest spirit priest of one hamlet in Dilli.[30] The number of *dungala* placed on the grave signifies his former rank in the Ngala society. At the end of the year, Ngala members gather at the grave site—each carrying his gourd emblem—to celebrate the final departure of the ancestral soul. The grave itself is then destroyed.[31]

It should be noted that during the preburial festivities held by the Mbula for a high-ranking elder, a plain calabash bowl is placed on the funeral bier along with the corpse. After burial, the gourd is moved to a forked stand (*kul*) erected alongside the grave. Beer and other offerings are regularly placed inside the bowl during the year the deceased is honored, serving much the same purpose as the small red calabash on the Ga'anda *hlefenda* vessel.

Divination is another context in which gourds play a crucial ancillary role. A number of groups in the region use calabash bowls as divining instruments. Among the Yungur, the diviner (*sife*) uses a plain gourd (*ding pengpeng*) sealed with red ochre to hold water into which various ingredients have been placed to call forth spirits associated with the proceedings (Fig. 33). Essentially, the *sife* asks a series of yes/no questions that are answered by the position of a small gazelle horn dropped into the water—a vertical position (seen in the photograph) is affirmative and a horizontal position is negative.[32]

Mwona diviners (*felan*) also use a gourd bowl to

hold a liquid medium for determining the causes of illness. Like the Yungur, the introduction of various objects into the liquid ritually activates them. By touching the places on the body where symptoms are felt with these items and then returning them to the liquid, the diviner reaches a diagnosis.

Other instances where gourds assist in divination were documented among the Pero (a Chadic group who live in the Benue-Gongola Valley) and the Burak (their Adamawa-speaking neighbors). Pero diviners (*ankwandul*) use two wood figurines (*kwandul*) to help determine the cause of illness; the Burak call the same type of figurines *kapgonol* (Fig. 34). The male/female pair each has an iron spike in its base and is embedded in the ground wherever the diviner is called to work. The figures are rubbed with red ochre and further activated through offerings of water, beer, and/or sorghum placed in tiny calabash bowls wedged directly into their mouths.[33] The intervention of the two wood figures in decision-making is supported by the diviner's name, *ankwandul*, "one who masters the *kwandul*." Actual divining takes one of two forms. In one, a gourd bowl is employed and, as in the preceeding examples, it serves as a tool that is essential for manipulating the primary divining device. The medium in this case is a small animal skin (*shiru*) with the head attached, studded with abrus seeds and pierced by an iron ring from which two bent-iron bells are suspended (Fig. 35). One end of the skin is tied in a loop and placed around the diviner's toe, while the *shiru*'s head rests on the edge of the gourd. The animal then dances up and down "on its own volition," jingling vigorously, until the diviner has arrived at the correct answer.

Two other divination procedures have been recorded that employ bottle-shaped gourds as primary instruments.[34] One is practiced by the Tula, a group who live in the hills northeast of the Pero. The diviner uses two gourds, one of which is a sprinkler filled with water and sealed with wax; and a second that is used as a rattle and contains seeds (Fig. 36). To initiate the divination procedure, the bottom of the water gourd is tapped against the rattle. While questions are being asked, the diviner inverts the sprinkler and, depending on how the streams of water are configured, determines whether an affirmative or negative answer is indicated. The rattle is used to announce that a final diagnosis has been reached.

32. Display (*tankul*) of ritual paraphernalia owned by a Mbula earth priest, including large gourd bottles (*dungala*) that signify membership in the Ngala secret society; Dilli.
Other displayed objects, such as bits of bicycle chain, keys, and iron bracelets, all contribute to the Menzali's ritual authority over things "found" in the earth. January 1982.

33. Yungur diviner (*sife*) using a plain gourd bowl as an instrument; Dirma. Note the vertical position of the small gazelle horn in the water indicating an affirmative answer to the diviner's question. May 1981.

34. Burak wooden divination figurines (*kapgonol*) with tiny gourd cups wedged in their mouths; Burak. March 1982.

35. Pero diviner (*ankwandul*) demonstrating the position of the divining apparatus—an animal skin (*shiru*) studded with abrus seeds and cowries on the edge of a gourd bowl; Gwandum. March 1982.

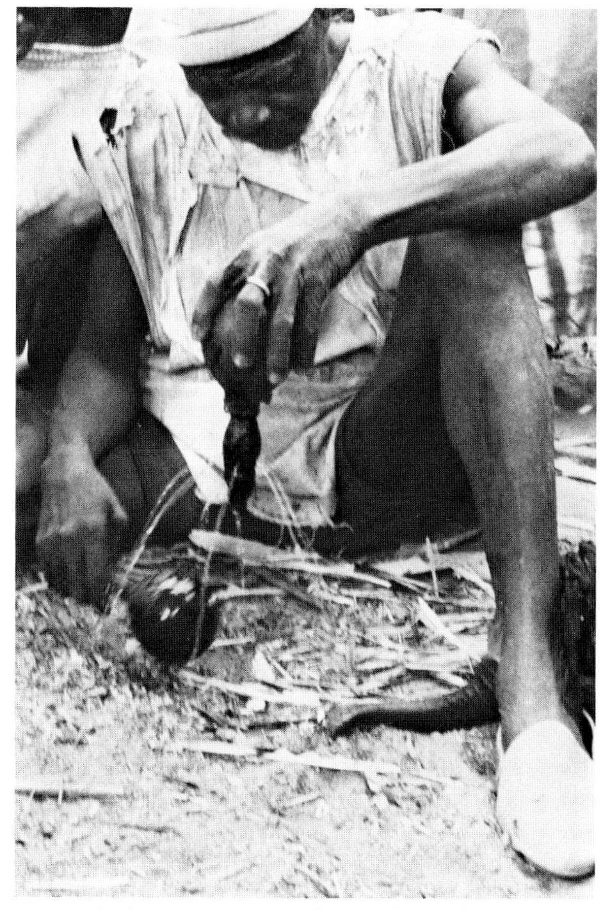

36. Tula diviner using gourd sprinkler to determine answers to his questions; Tula Yiri. 1970.

37. Bura leather-covered bottle gourd (*jatau*) suspended on a string during divination; Pelambirni. 1970.

The second divination procedure employing a bottle-shaped gourd was recorded among the Bura. A gourd covered with strips of leather and cowrie shells (*jatau*) is suspended on a cord tied at one end to the diviner's toe and held at the other in his hand (Fig. 37). Questions are asked and if the gourd slips down the cord to the ground—i.e., tension is increased—an answer is indicated.

One final example of divination was documented among the Mumuye who push a long tubular gourd into a deep elliptical bowl filled with water to create a "voiced" oracle (pc:A. Rubin, December 1984). The long flaring gourd is particularly distinctive as it is bound with rope while still on the vine to create deep grooves in its surface. Near the mouth, spider egg sacs are applied with gum so that when the tube is forced into the water, the air rushing out causes the gourd to "speak" with an eerie whistle.

The use of gourds in divination may relate to the symbolic meanings that have been associated with these containers. For example, Chappel argues that for the Yungur a plain gourd "has the power to restore the status quo whenever the social situation has been, or is about to be, dangerously disturbed" (1977:18). Illness or accusations of witchcraft certainly represent such disturbances and thus make gourds appropriate symbolic as well as instrumental components of divination.[35] Berns was told by Yungur informants that two disputing villages could arrange a truce if a "white" (i.e., plain) gourd was overturned at the crossroads between them. Chappel notes that the "simple expedient of putting down a plain gourd" extends even to disputes that have reached serious proportions in a household, lineage, or village (1977:20). He stresses the symbolic link between a "white gourd" (*ding pengpeng*) and its power to "cool" or, literally translated, "to make white" (*kal pengpeng*) a person (1977:18–20). Placing a gourd between two disputants cools them so their anger will be dispelled. According to Chappel, a plain gourd is given by parents to the woman who will perform scarification on their daughter. It acts as a cooling agent and ensures that the woman cutting the marks has the patience to do her work well. While these associations between relieving situations of stress and the presence of a plain gourd seem reasonable, Chappel's contention that the essential shape and botanical structure of gourds explain why they have become expressive equivalents for the stability and order to which societies strive may be stretching the argument too far (1977:16–18). Rather than a response to the physical regularity and symmetry of this fruit with features like "a mass concentrated around its centre" or "the balanced 'pull' of the dynamic forces irradiating in all directions from a center which, far from being 'dead,' is alive with tension" (Chappel 1977:16), the translation of the gourd into a primary symbol of social integration may more likely derive from its fundamental economic utility, a utility that ultimately does depend on a respect for the gourd's inherent structure. What was suggested earlier based on the observations of David and Hennig should be repeated: gourds were not likely to have become an important focus of ritual and, by extension, symbolic thought unless they already were highly valued in domestic and social spheres (1972:28).

This point is supported by the ways a gourd is used as a vehicle for symbolic verbal expression. The Hausa proverbs presented earlier are clear examples of this application in common speech. Additionally, the nickname purportedly coined by the settled Fulani for the "pagan" groups living in the Ga'anda Hills—the Ga'anda, Gbinna, and Yungur—is "Lala." Temple defines this word as "an old calabash broken into many parts," referring in a derogatory sense to the scattered and decentralized populations of this rugged area (1919:255). This definition implies that the Fulani regarded such settlement and organizational patterns with disdain, having as little utility as a broken gourd. The Kanuri, who also saw themselves as culturally superior, apparently gave one of the groups living around Potiskum its name—Karekare. According to Meek, the name derived from the composite character of the group, which resembled "a patched-up calabash" (*wasak kere*; 1931 II:220). Whether the origins of these ethnic designations and their meanings can be attributed to Fulani or Kanuri sources remains equivocal. However, what continues to be important is how gourds have been used, both actually and metaphorically, as expressive devices.

MUSICAL INSTRUMENTS IN SOCIAL AND SACRED CONTEXTS

Gourds, both decorated and undecorated, figure in various public festival contexts not only as ornaments and symbolic attributes, but also as musical instruments. During sacred or ceremonial events, gourd instruments can create distinctive tonal "voices" that link the community to the forces regulating its survival. One of the most notable examples is the large drum (*dimkedim*) played at the Wora funeral celebration held by the Yungur (Pl. 14). Two large globular gourds, joined together with rope covered in dung, create a lower resonating chamber and are attached to a hollow wood cylinder fitted with an antelope (duiker) skin head. Only at such funerals are three of these spectacular drums played together by male members of one royal lineage (Bera Kumla; Fig. 38). The origin of this drum configuration is not known, although the Yungur claim the first set of *dimkedim* was brought by early migrants from their ancestral homeland, a hill named Mukan. This ascribed provenance historically legitimizes these drums and associates them with powerful ancestral forces that are traced to this site and upon whom the Yungur depend for survival. Beating these drums at funerals may be a way of establishing contact with the spirits of family ancestors, including the recently deceased, who are collectively honored on such occasions.

A related tradition is maintained by one Longuda subgroup, the Dumna Zerbu, who use a calabash wind instrument (*sumeh*) to communicate with powerful supernatural forces (Fig. 39). The *sumeh* is made from one large spherical gourd to which a long curved tubular extension is attached with gum and rope. This trumpetlike instrument is offered sacrifices to ensure the efficacy of its voice. It is blown at dances held at the end of the farming season to thank the forces responsible for the successful harvest. It is heard again during festivities that inaugurate and promote the success of dry season hunts. The *sumeh* used on such occasions is kept in the village and is replaced if broken. It is regarded as a copy of the first instrument given to the Dumna Zerbu by their spirit guardians; the original's sacred voice is heard each time an elder dies, emanating from the nearby hills where it is thought to be enshrined. This suggests a certain consistency in the association of resonating gourd voices with the transition of the living to the realm of the dead.

In Bodwai, a Gbinna village located near the Gongola-Hawal confluence, an unusual gourd trumpet (*muhamra*) accompanies a band of xylophones (*shilan'ja*) at funerals (Fig. 40). It is approximately 1.5 meters long and is made from four tubular gourd sections, each one fitting into the other, increasing in diameter from the mouthpiece downward. The joints are rendered airtight with an application of the rubbery substance produced by sweat flies.[36] Similarly constructed and contextually related gourd trumpets, called *vadõsõ*, are used by the Mumuye in Vabõ masquerade performances associated with collective funeral celebrations ("all souls festivals"; Fig. 41).[37] The trumpets' voices evoke the presence of ancestral spirits who are called forth during such celebrations.

38. Yungur men playing a *dimkedim* during Wora; Dirma. November 1981.

39. Longuda gourd trumpet (*sumeh*) played during harvest and hunting festivals; Dumna Zerbu. November 1981.

40. Demonstration of Gbinna musical instruments, including a gourd trumpet (*muhamra*), bamboo flute (*b'eah*), pottery drum (*dara*), and double-headed log drum (*bombilara*); Bodwai. December 1981.

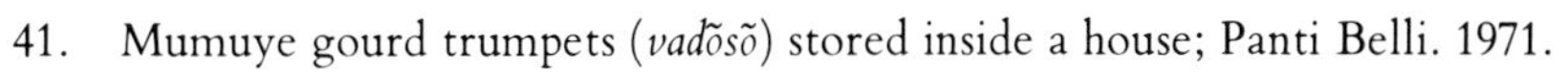

41. Mumuye gourd trumpets (*vaɗõsõ*) stored inside a house; Panti Belli. 1971.

The Ga'anda also associate gourd voices with the ancestors. Five to seven fixed-key xylophones (*kila-ya*) are played together at funeral ceremonies in the presence of the corpse laid out on a wood bier. Only one Ga'anda family maintains the right to own, construct, and play these xylophones. At one time cylindrical gourd resonators of graduated length were set under seven hardwood keys suspended in a wood hoop; today, cow horns have been substituted. Each instrument is played with a pair of forked wood sticks, a configuration that allows the musician to strike up to four keys at once.[38]

The distribution of gourd xylophones extends to other areas of Nigeria, as well as far beyond its western and eastern borders.[39] One geographically proximate Nigerian example was documented among the Chamba whose xylophone has thirteen fixed keys, each with a gourd resonator of gradually increasing size (Fig. 42).[40] Small holes are pierced near the ends of the gourds so that spider egg sacs can be secured across them with beeswax. This produces a slight buzzing sound when the instrument is played.

The preceding examples emphasize the importance of gourd instruments in specific ritual contexts where their ownership and performance are often the exclusive prerogative of special families. Calabashes are also used to make a variety of instruments played by professional musicians in northern Nigeria, as well as across the continent.[41] The gourd has long been adapted to musical use because of its versatility as a resonating chamber that can be struck, blown into, or scraped. The Hausa, who consider musicianship a traditional craft, make a great variety of instruments from calabashes of different shapes and sizes. Hemispherical bowls can be overturned and struck with the fingers (with or without rings) or with sticks (*k'warya*); bottle gourds can be filled with stones and used as rattles (*caki*; Pl. 15); globular gourds ranging from 30 to 70 cm in diameter are made into kettle drums (*gwazogwazo*, *duma*, *dumuniya*, *talle*); five varieties of spike-bowl lutes with one or two strings plucked or played with a bow are made with calabash sound boxes over which skin has been stretched (*goge*, *gurmi*, *kukuma*, *garaya*, *kuntugi*); and circular pieces of gourd shell threaded onto a cord are used as a sistrum (*kwak kwafa*).[42] A long tubular gourd is also used as a percussion instrument (*shantu*), played exclusively by women.[43]

42. Chamba gourd xylophone (*tikari*) being played in Takum. 1970.

The performer generally sits cross-legged and with her left hand taps the calabash against her thigh, calf, open palm, shin bone, or the ground, while her right hand controls the various sounds by intermittently closing off the upper end (Ames and King 1971:10–11). Most *shantu*, also made by the Kanuri and Fulani after Hausa models, are decorated along their entire length (Fig. 43). Beyond a purely musical role, the sounds emitted by this instrument are used by Kanuri women to imitate sequences of syllables in their language (Wente-Lukas 1977a:261). By playing them, women are thus able to convey to their female relatives and friends messages not understood by men, whom they mock in this fashion.

This comparative ethnography of gourd use provides a basis for understanding why this material has become the object of such intense artistic elaboration, both in this region and virtually across the continent. Yet, the ethnography also shows that there are certain ceremonial and sacred contexts where gourds are left unembellished. It is likely that these plain gourds, which figure prominently in activities dominated by men, lack the programs of decoration usually executed by and associated with women. As indicated earlier, certain ritual priests are forbidden to eat from the ornamented gourd bowls typically used by women to serve their husbands food. Gourd instruments played in ceremonial contexts by men are usually left unadorned. It is telling that the *shantu* drum played exclusively by women is elaborately decorated (Fig. 43). While a multiplicity of gourds are essential to life in northeastern Nigeria, those that are the most highly decorated are generally used and displayed by women.

In sum, the time and labor invested in gourd decoration enhance the economic and social implications of the fruit's cultivation and use. Because they are made "better to look at," decorated gourds also are regarded as "better to use."[44] The display contexts described above indicate that among many groups, calabashes are intended to be viewed and admired. Moreover, it may be that because they satisfy an essential material need, their decoration is a socially sanctioned act. The ability to decorate them well gives women, and some men, a special avenue for achieving social recognition and prestige. The number of techniques used to transform gourds and the diverse combinations of designs that have been worked over their surfaces testify to the vitality and value of this artistic enterprise.

43. Kanuri percussion instrument (*shantu*). 59.6 cm. UCLA MCH X85–55.

·3·

AN ETHNOGRAPHY OF GOURD DECORATION

Techniques, Designs and their Meaning

TECHNIQUES

Gourds have hard yellow shells a few millimeters thick, protected by a green outer skin usually removed before decoration, and softer white inner-layers that vary in thickness. The surfaces of most gourds, regardless of their shape, are amenable to decoration, and differential color and porosity have influenced the evolution and application of decorative processes. The exterior texture of tubercled gourds is considered a natural form of decoration (Fig. 44); they can, however, be ornamented on the interior.

Like the numerous shapes and functions of the gourd, the options for artistic embellishment are many and permit the same degree of innovation and versatility. Beyond the imposition of design and pattern, the gourd lends itself well to changes in color, texture, and relief. Even if no ornamentation is applied, the surface changes with time. Through handling, saturation with butter or other oils, and exposure to smoke, gourds develop a rich brownish-red or dark honey-colored patina. As Sieber has observed, a "well-used, well-loved calabash may display a marvelous combination of form, decoration, age, patina, and repair" (1980:195).

Northeastern Nigerian peoples use six techniques to decorate gourds: 1) pyro-engraving; 2) pressure-engraving; 3) carving (scraping); 4) painting; 5) dyeing; and 6) adding decorative materials (Map C). Each group's calabashes are decorated with one primary technique, although supplementary processes are often combined for added effect and greater individuation.

Pyro-engraving involves burning lines into the surface of the gourd with a hot metal blade and is the most widely used technique.[1] It is found among the Dera, Pidlimndi, Tera, Bura, Pabir, Kilba, Chibak, Margi, Mumuye, Dadiya, Kanuri, Bata, Mboi, Mbula, Waja, Yungur, Hausa, and settled Fulani. Pressure-engraving involves drawing a sharp point across the gourd surface while applying a considerable amount of pressure. It is practiced by groups who live in a fairly circumscribed geographical area—the Ga'anda, Gbinna, Hona, Yungur, and Longuda—as well as by the pastoral Fulani who are widely distributed across northern Nigeria. Carving involves outlining designs with a sharp blade and then cutting away the outer cuticle. The remaining cuticle of the gourd stands out in relief against the reduced background. The Hausa are the only people represented in this collection who use this technique as a primary decorative device; the eastern Margi also use it, but only in combination with oil-resist dyeing. Other groups use minimal carving to create relief variation on gourds decorated with a different primary mode. The fourth technique, painting, was traditionally confined to the interior gourd surface. Its distribution as an exclusive ornamental procedure is fairly restricted, limited to peoples living in the Potiskum Plains—the Karekare, Ngizim, and Ngamo. Although rare, the settled Fulani also paint the inside of gourds pyro-engraved on the outside. In addition to its decorative function, painting seals the semiporous gourd with oil-based pigments to better hold liquids. Dyeing the outer shell is usually used in conjunction with other decorative processes. The settled Fulani, eastern Margi, Bata, and Ga'anda are the main groups who do this. Lastly, a wide range of decorative materials can be added to the surface of the gourd. The embroidered letters on the Mwona gourd and the iron handle laced onto the Dadiya dance rattle are only two examples (Figs. 20,26). This additive technique is less evident in northeastern Nigeria than it is in other areas of Africa where gourds are decorated with materials like leather, beads, basketry, or plaited raffia. Prestigious materials such as glass beads, cowrie shells, and metal wire (including brass, steel, and gold) are also used.[2]

Before a thorough explanation of these techniques is given, it should be stressed that the exploitation of each decorative process varies considerably from group to group. Supplementary procedures further extend the range of diversification, and how the decoration is physically executed becomes a determining factor in the evolution of distinctive design programs. Technique and design have their own independent variables, and developments in composition depend on the harmonious integration of both factors.

What should be understood, in light of the distribution, range, and number of techniques used in northeastern Nigeria, is that choices made by individual groups are based on more than just considerations of aesthetic values and criteria. Choices of technique and design are conditioned by various sociohistorical realities. The diversity of calabash decoration in northeastern Nigeria is inseparable from its geographic, linguistic, and ethnographic complexity.[3]

Pyro-engraving The forged iron tool used in pyro-engraving generally has a blade shaped like an arrowhead with a long, straight shaft embedded in a circular wood handle (Fig. 45a). Artists work with one or more such tools (usually three to six), all of them heated in a fire so that as the one being used cools another is ready to replace it (Pl. 16). The fire is stoked with thick branches, sometimes dampened, to ensure they burn slowly and without creating enough heat to deform the blades. Artists use the tools in order of their placement in the fire. Before (and sometimes after) each knife is used, it is scraped against a stone to remove the carbon residue. The Fulani knock the tool against a piece of wood or jab it quickly into the sandy soil (Chappel 1977:36). Artists may also plunge the knife into a bowl of water to clean the blade.

To execute a line, the carving hand, separated from the gourd by the length of the tool, moves the knife toward the body while the other hand

44. Tubercled gourd. 17.5 cm x 18.5 cm. Russell and Maxine Schuh collection.

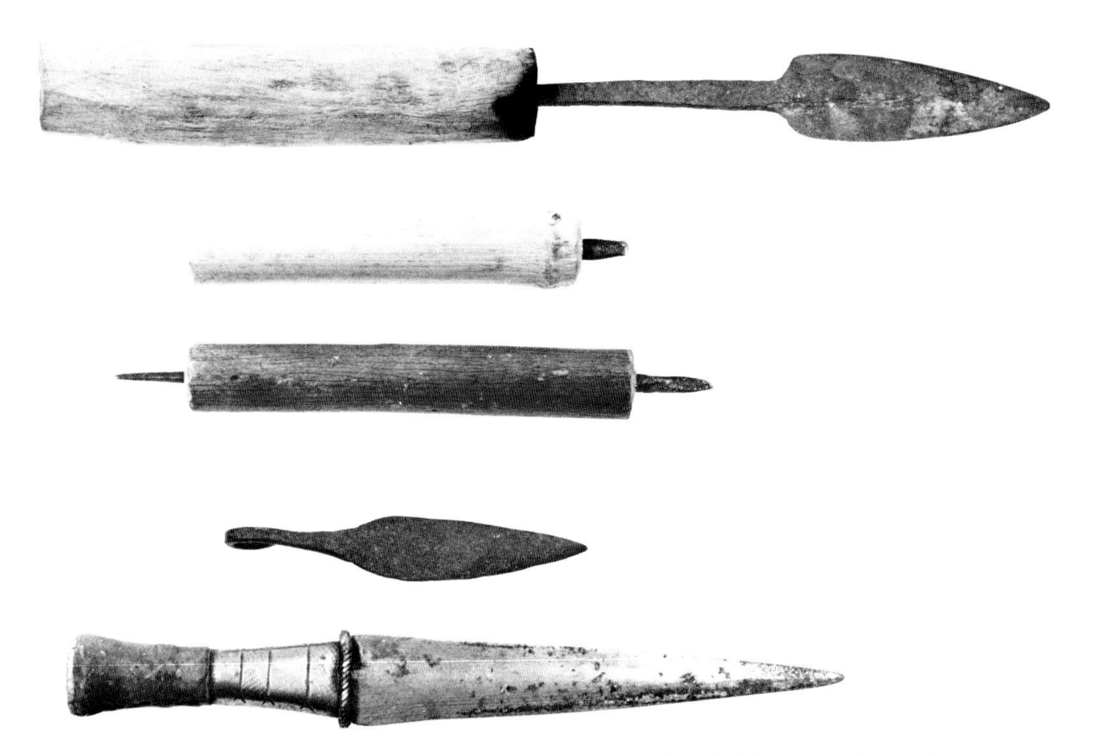

45. Decorating tools. Top to bottom: a. Tera pyro-engraving tool (*sud'ida*). Iron and wood. 33.6 cm. UCLA MCH X83–814; b. Ga'anda pressure-engraving tool (*xafta*). Iron and guinea corn stalk. 14 cm. UCLA MCH X83–817; c. Gbinna pressure-engraving tool (*b'aa kõsa*). Iron and wood. 19 cm. UCLA MCH X85–62; d. Longuda pressure-engraving tool. Iron. 11.4 cm. UCLA MCH X83–818; e. Hausa carving knife (*wuk'a*). Iron, wood, and leather. 23.5 cm. UCLA MCH X83–816.

rotates the gourd, held sideways or in an inverted postion, in the opposite direction. Chappel observes that it often appears as if the knife is kept stationary and that the black engraved lines result from turning the gourd (1977:36). In working lines perpendicular to the edge, for example, the artist may hold the gourd by the rim moving it upward like a lever against the knife blade.

The basic technique used in pyrography is essentially the same everywhere, but variations are achieved by altering the shape of the knife, the area of the blade used for burning, and the amount of pressure applied. For example, the Tera use the sharp edge of a leaf-shaped knife with moderate pressure to burn-in the tight feathery lines (*ndesa*) that characterize one aspect of their engraving style (Fig. 46). The Dera, on the other hand, apply slightly more pressure to achieve deeper and somewhat broader incisions. This rather subtle difference in texture is compounded by a secondary process—rubbing a soft stone over the unmarked areas so that the abraded cuticle appears much lighter than the burned-in lines (see Pl. 26; Fig. 75).

A distinctive variation in knife handling is found among the Bata and Mboi who apply great pressure to the broad surface of the blade. This creates deep heavy lines that have a U-shaped contour, rather than the V-shape that characterizes the cross-section of most engraved lines (Fig. 47).[4] A similar effect is achieved in a highly individual gourd engraved by a Kanuri artist (Pl. 28b). Again, the broad section of the knife blade was used with considerable pressure to draw an elaborate pattern of deeply incised lines. What especially distinguishes this Kanuri gourd, however, is the use of the knife tip to impress a continuous frame of small deep cuts around the lines.[5] This use of the knife is an example of how technical inventiveness can enhance an already vigorous composition.

The use of a very fine blade is unique to a number of groups living in the Uba Plains—the western Margi, the Kilba, and the Chibak (Pl. 28a; Fig. 48). Delicate, closely placed lines burned-in with a minimum of pressure are a strong textural counterpoint to the broad bands of deeply engraved lines that form a geometric framework across the gourd surface.

The Mbula produce an unusual color and textural variation by allowing the knife to cool partially and then wiggling it back and forth over the surface. In this way a distinctive pattern of reddish-bronze wavy lines is achieved (Fig. 49).[6] The Kilba

46. Detail of "feathery" lines used in Tera engraving (*ndesa*). 24.1 cm x 23.5 cm. UCLA MCH X83–687.

47. Detail of the "heavy" burning technique used by the Bata. 27 cm x 31.1 cm. UCLA MCH X83–673.

48. Detail of "fine line" cross-hatching used by the Kilba. 25.5 cm x 30.2 cm. UCLA MCH X83–763.

49. Detail of "rocking" technique used by the Mbula. 20.3 cm x 21.6 cm. UCLA MCH X85–43.

sometimes use the same "rocking" technique to lightly burn the fine hatched lines that characterize the filled-in areas of their gourds (cf. Fig. 48). These dark brown lines create zones of intermediary color and texture, which in both cases contrast with the heavy black engraving and bands of unworked yellow shell.

Still another effect is produced by scorching broad areas of the gourd with the flat surface of the blade. This procedure is often employed as a framing device since it is easy to manipulate the blade near the rim. It is also used to blacken the cut edges of thick-walled gourds. A number of groups have exploited this method to achieve other decorative results. The Mbula use the flat edge of the blade to scorch over broad bands of densely hatched lines and thereby achieve a bolder contrast between worked and unworked areas (Fig. 50). The Kanuri also use scorching to fill in shapes that then stand out boldly against unworked ground (see Figs. 86,87). Design registers dominated by forms that have almost been "painted" with the side and flat tip of the knife blade provide a striking counterpoint to areas that have been cross-hatched using broad sweeps of the knife edge.[7] The Kanuri gourd also shows how a wide band of rim scorching is used effectively to frame an elaborately engraved composition. In other areas of Africa, this scorching technique is used to create impressive patterns through shape and color contrast rather than through textural variation.[8]

In contrast, a light application of the knife blade is evident in the way the interior surfaces of some calabashes are decorated (Fig. 51). This procedure is generally restricted to gourds with thick inner walls that have a soft surface that is easily burned. The ability of some artists to achieve delicate and closely placed lines within the confined concave space of the gourd interior demonstrates great control and dexterity (Fig. 51a).

Another remarkable example of engraving skill is evident in the gourd "sun bonnet" attributed to a Yungur carver (Fig. 52). In this case, however, the artist has used a heavy burning technique like that of the Bata (cf. Fig. 47).

Pressure-engraving The tool used in pressure-engraving is generally an iron point embedded in a sorghum stalk or in a wood handle (Fig. 45b,c). The gourd is held either sideways or in an inverted position, usually resting on the thigh or against the stomach and chest. As the tool is pulled perpendicularly across the surface with the right hand, the left hand keeps the gourd firmly in position (often pressed tightly against the body) and then rotates it slightly as the work progresses (Fig. 53). For most artists, the key to effective engraving seems to be in bending over the gourd so that the weight of the body is behind the arm movements. Chappel provides a good description of how the technique is executed:

> Engraving is a laborious and painstaking activity requiring the application of a fair amount of physical pressure as the engraving tool, gripped like a dagger, is dragged across the surface of the gourd. Arm movements, essentially, are involved, for the wrist is locked in a rigid or semi-rigid position, while the carving hand is itself in direct contact with the working surface of the gourd . . . The technique favours concentration on small areas of the design field, the position of the gourd being frequently adjusted to allow for the maximum degree of controlled pressure on the engraving tool (1977:34).

All groups in this study employ pressure-engraving in two distinct stages. In the first, lines are incised across the surface to define the overall composition (Fig. 53). Although linear patterns sometimes suffice on their own, normally, in the second stage, spaces between parallel lines are delicately hatched or entirely scraped away, often in alternating bands. The second stage is the more laborious of the two, and with it, the distinctive and variable textural effects of this reductive process are achieved. The incisions made during both stages expose the white underlayer of the gourd's shell. But as Figure 54 reveals, very little of the surface material is removed in making the delicate incisions, and the decorative patterns are barely visible except at close range. Therefore, the engraved design is usually filled in with a blackening agent allowing the "reversed" patterns to stand out crisply against the unmarked ground.

The Ga'anda call the underlying "drawn" composition *njoxta*, and it is generally comprised of shapes inscribed by a series of concentric lines and other repetitive geometric motifs. The designs and patterns are brought out by delicate cross-hatchings called *cerahla* that fill in the alternating spaces between lines (Fig. 54). The Ga'anda then rub a mixture of oil (*fideta*) and soot scraped off the bottom of a cooking pot (*mərkukwina*) over engraved areas of the gourd. Then the gourd is rubbed with dry earth, "removing the oily charcoal that is on the surface and leaving a residue only where lines have been pressed into the calabash" (Fig. 55; B. Rubin 1970:25).

50. Alaja Linga scorching the surface of a Mbula gourd bowl (*kwar*); Borrong. January 1982.

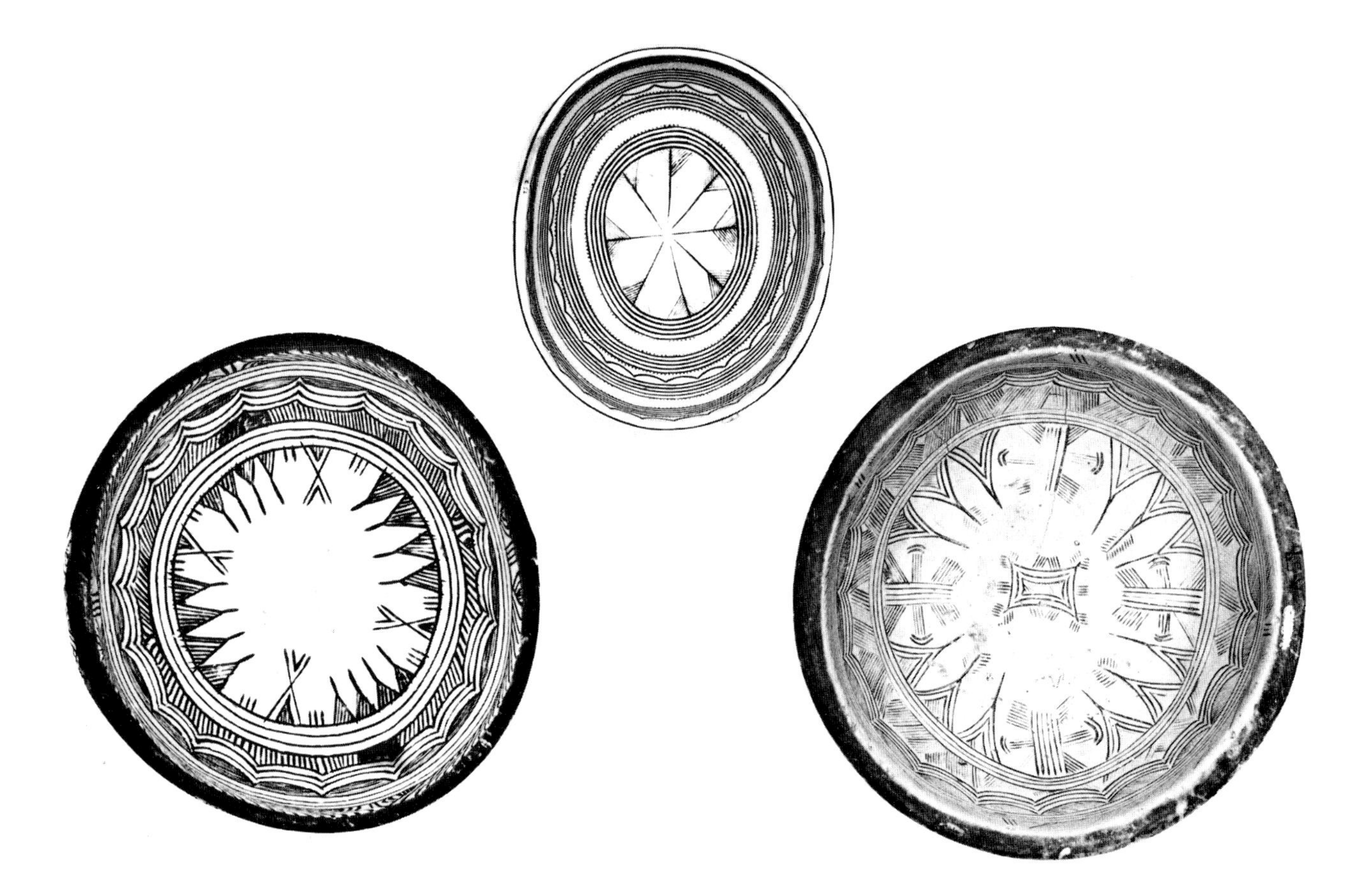

51. Interiors of gourds.
a. Dera (*lib'e tib'ekokumat*). 17 cm x 19.2 cm. Berns collection;
b. Tera (*d'eba*). 24 cm x 24.2 cm. Berns collection;
c. Tera. 25.4 cm x 25.7 cm. UCLA MCH X83–695.

52. Yungur gourd "sun bonnet." 26 cm x 29.5 cm. UCLA MCH X85–28.

The engraved lines produced by the Ga'anda are slightly thicker than those of the pastoral Fulani (Wodabe) who use a finer point. Their engraving tool, called a *jalbal*, is a multipurpose iron needle about 10–12 cm long (Chappel 1977:42). It is basically a narrow metal shaft with two pointed ends, one of which is beaten into a leaf shape similar to an unbarbed arrowhead. According to Chappel, this type of point was traditionally used for pressure-engraving and the *jalbal*, with its two points, is the modern equivalent (1977:42). A wood grip is fashioned to cover the center of the tool. While the *jalbal* is the tool most often used, it appears that any sharp instrument will do—one Fulani carver preferred to use a pointed valve from an old automobile (Chappel (1977:42). In the same way that the chisel-shaped point used by the Ga'anda crisply executes *njoxta* and *cerahla*, the use of a sharper blade is more suitable for the two stages of the Fulani method (*tuppuchi*): the design is drawn with rather fine lines and then the point is repeatedly scraped across the area "between alternate pairs of engraved lines until the surface shell has been removed, so that the white core is exposed" (Chappel 1977:42). The engraved areas are then filled with a mixture of butter and grasses burnt into ashes. Chappel writes that the process is often repeated at intervals to renew the darkening. He also describes a recent innovation that fixes the designs more effectively and for a longer period—a filling is made from butter and the acid extracted from old flashlight batteries (1977:42). The blackening adheres to the scraped areas, creating a striking pattern of marked and unmarked alternations, one recessed and one in relief (Fig. 56). This broad-groove engraving technique is distinctive to certain Wodabe groups even though a pattern of strict alternations also characterizes the pressure-engraved work of other peoples. The Ga'anda technique of using delicate cross-hatchings (*cerahla*) to fill between the lines creates quite a different effect than does entirely scraping them away (cf. Fig. 55).

The fine point on the Fulani *jalbal* needle also allows an artist to draw linear motifs that need not be scraped in between to achieve thicker lines or alternations. Single lines and dense hatchings can be engraved to create contrasting motifs and textures (Figs. 57,58). Sometimes the decoration is entirely the result of elaborate engraving, the surface left unblackened (Fig. 58). This variable use of the engraving tool creates graphic designs that are quite distinctive from the singular alternating grooves used by other pastoral Fulani groups.[9]

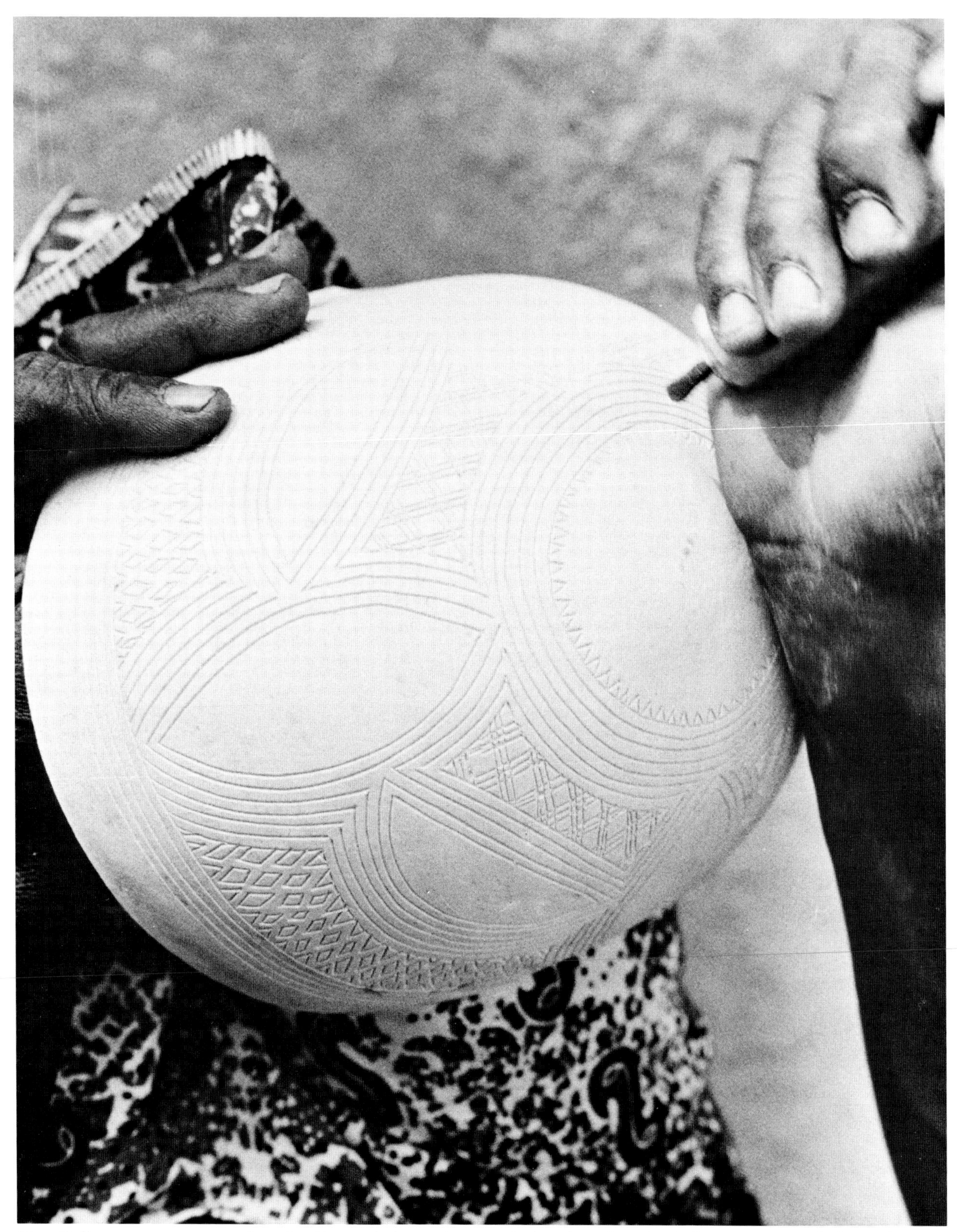

53. Kalhar pressure-engraving the linear design, *njoxta*, and showing the physical constraints of the technique; Ga'anda. January 1981.

54. Unblackened Ga'anda gourd bowl (*kowata*). 14.6 cm x 17.6 cm. UCLA MCH X85–1.

Gbinna women also decorate calabashes (*kõ*) with pressure-engraving (*yoroma*). Like the Ga'anda, they use a chisel-shaped iron point for cutting lines across the surface to mark out the design. In contrast to the Ga'anda who rely on a dense network of cross-hatching to create texture, the Gbinna gouge out portions of the gourd surface with an engraving tool (*b'aa yor kõjeba* or *b'aa kõsa*) that has a second awl-shaped blade embedded into the opposite end of the handle (Fig. 45c). Thus, the surfaces of Gbinna calabashes tend to be more sculptural, and filling the carved, recessed areas with oil and charcoal intensifies the contrast between textured and untextured areas (Fig. 59).

The Hill Longuda, unlike any of their neighbors, make use of the outer cuticle that protects the hard shell of the gourd. They impress lines directly into the soft green surface using a small curved blade or some other sharp instrument (Fig. 45d) and then bring out the patterns by applying a mixture of oil and charcoal to the surface (Fig. 60). As the cuticle dries, it changes from green to warm yellow. Hopefully, the cuticle continues to adhere to the gourd shell as it dries. Over time and with use, however, this outer skin does wear away, taking the design with it. Longuda women claim that gourds keep their designs for at least several years.

While this would seem to be an eccentric way to decorate gourds, the Longuda technique is not as laborious as other modes. Impressing a design into the soft cuticle requires less physical pressure than the Hona, Ga'anda, or Gbinna method of engraving or even the Tera and Bura method of pyrogravure. The technique does, however, require considerable patience and control to achieve the delicate, subtle all-over designs characteristic of the Longuda engraving style (Fig. 60). The inevitable cycle of design and decay means there will be a constant demand for the skills of the artist.

Carving Carving a gourd creates the reverse effect of pressure-engraving—portions that are scraped become the background, whereas areas that stand in relief form the design (Fig. 61a,c). The subtle pattern that emerges as yellow against white is often intensified by the addition of white chalk or clay to the scraped ground and/or stain to the remaining relief (Fig. 61b). Sometimes pyro-engraving is combined with chip-carving (Fig. 62a); the interiors of such gourds may also be pyro-engraved (Fig. 62b). The carving tool has a blade that is wider and longer than the point used in pressure-engraving, which allows more of the hard yellow shell to be removed (Fig. 45e). Very

55. Detail of Ga'anda pressure-engraving. 20.6 cm x 25.4 cm. UCLA MCH X83–619.

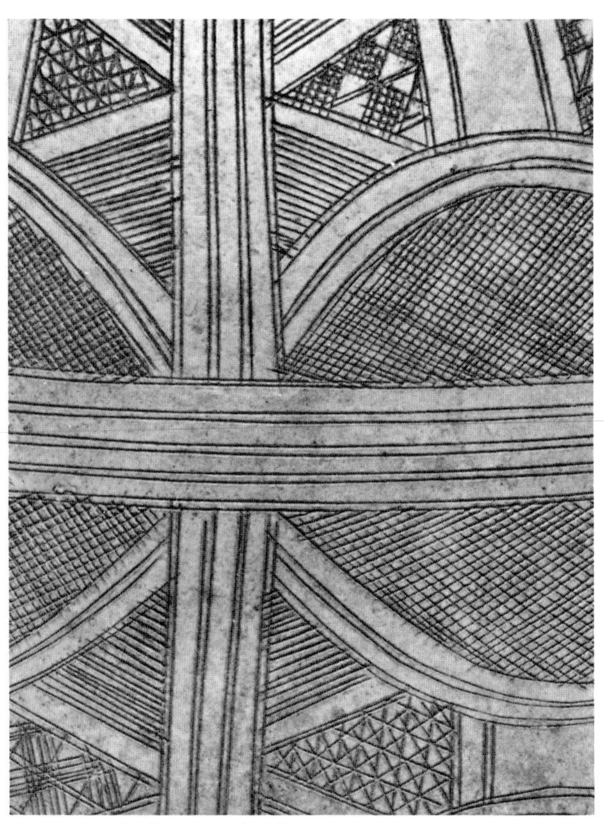

58. Detail of pastoral Fulani pressure-engraving of the exterior of a gourd decorated on the interior by a Kare-kare woman. 39.4 cm x 41.9 cm. UCLA MCH X83–779.

56. Detail of pastoral Fulani pressure-engraving. 27.3 cm x 28.9 cm. UCLA MCH X83–784.

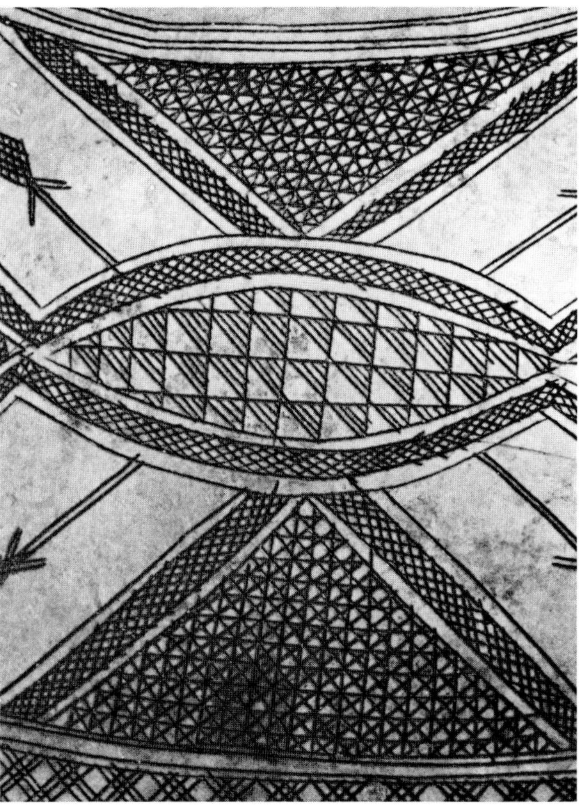

57. Detail of pastoral Fulani "style" pressure-engraving executed by an Ngamo woman. 19.0 cm x 20.6 cm. UCLA MCH X83–775.

59. Detail of Gbinna pressure-engraving. 20.9 cm x 21.1 cm. UCLA MCH X85–35.

60. Detail of Longuda pressure-engraving. 22.2 cm x 22.5 cm. UCLA MCH X83–658.

61. Hausa carved gourds.
a. Gourd bottle (*buta*). Leather. 42.4 cm x 24.1 cm. UCLA MCH X83–794;
b. Gourd bowl (*k'warya*). Clay. 21.9 cm x 22.2 cm. UCLA MCH X83–790;
c. Gourd bowl (*k'warya*). 37 cm x 34.5 cm. Russell and Maxine Schuh collection.

62. Hausa gourd.
a. Exterior view; b. Interior view.
28.5 cm x 28.8 cm. UCLA MCH X66–914.

deep carving, possible on gourds with extra thick walls, produces quite different results (Fig. 63).[10]

Whereas all procedures described above are predominantly used by women, most calabash decoration done by the Hausa is the work of men.[11] This is especially true of gourds ornamented by chip carving, and it is common to see groups of men decorating calabashes for sale at local markets. Along with decorated gourds, Hausa men also sell plain prepared gourds that they cultivate in large quantities. In many areas, they supply the Fulani with the deep bowls used to hold milk. These containers are often purchased already carved and finished with white clay rubbed over the dampened scraped portions of the gourd surface (Heathcote 1976:46). Other Fulani patrons favor gourds that have had the raised shell stained with a reddish-brown dye to intensify contrast.[12] One calabash bowl of this type, collected among the Fulani, provides a good example of Hausa chip carving (Fig. 61b).

Some groups engrave elaborate designs on the raised panels of hard cuticle. The smooth, uncarved surface may then be lightly stained or left to develop a rich patina, thus making relief designs visible in reverse fashion to the filling-in of pressure-engraved incisions.[13]

A carving technique used by the Tera, Waja, Bura, and Jera entails removing a 1–3 cm wide band of the gourd cuticle from around the rim. This not only provides a thin handling edge but also frames the design not unlike the rim-scorching used by other groups. Both the Bura and the Waja carve one or more deep grooves around the gourd parallel to the cut edge leaving pyro-engraved strips in between (Fig. 64). Sometimes the Bura and Pabir carve a series of triangular wedges around the gourd, starting from the rim end and pointing toward the center (Fig. 65). This supplementary decorative procedure achieves a three-way contrast in the color and texture of the gourd—the lustrous patina that develops on the unmarked shell, the coarser lighter-colored carved portions, and the black engraved linear patterns. White chalk may be rubbed into the portions carved away to intensify color contrasts.

While the eastern Margi use some pyro-engraving in decorating their gourds, they also employ a unique carving technique to exploit the soft, green cuticle on the outside of the gourd. Ordinarily, this outer shell is scraped away prior to decoration, but here it is used as a temporary stencil. Only the Hill Longuda, described above, similarly exploit the ephemeral outer skin of the gourd in

63. Hausa gourd. a. Bowl. 28 cm. UCLA MCH X65–5241; b. Lid. 26.4 cm. UCLA MCH X65–5240.

64. Detail of Waja gourds (*kamwone*) with carved grooves. a. 23.6 cm x 24.6 cm. UCLA MCH X85–51; b. 24.7 cm x 25.4 cm. UCLA MCH X83–717.

65. Gourds with carved "wedges" (*feliya*). a. Bura. 29.5 cm x 29.2 cm. UCLA MCH X83–686; b. Pabir. 23.2 cm x 23.5 cm. UCLA MCH X83–712.

the process of pressure-engraving. Yet, while the Longuda incise their designs directly into the cuticle with the awareness that the epidermis (and consequently the design) can easily wear away (evident in examples in the collection; Fig. 70), the Margi technique is combined with resist-dyeing (*kabaki*) and therefore lasts longer.

An uncut gourd selected by a Margi artist for decoration is chosen as much for the intactness of its cuticle as for the appropriateness of its size and shape. The outer skin of an uncut gourd is carefully incised in a geometric pattern, and the portions to be colored are carefully scraped away with a triangular razor or other sharp tool (Fig. 66). The scraped areas are then rubbed with shea nut oil and the whole gourd is submerged in a pot of boiling water for several hours until the oiled portions have turned red.[14] When the gourd is cool, what remains of the cuticle is shaved off, revealing a contrasting ground of yellow—the natural color of the shell of the gourd. Because of shrinkage and/or the cuticle having peeled away during the boiling process, the dyed lines spread, leaving not the sharp geometric pattern that was cut into the surface, but a softer, broader, somewhat fuzzy version of the original design (Pl. 18b). Because gourds are boiled whole, the size of Margi-dyed gourds is limited to those that can fit into clay cooking pots.

66. Margi woman scraping away green, outer cuticle of the gourd; Gulak. 1971.

Painting The Karekare, who live ten miles south of Potiskum on the road to Bauchi, process the raw materials that they and the Ngamo and Ngizim use to decorate gourds. The interior surfaces are painted with pigments extracted from clay and from other substances. This decorative procedure depends on the exploitation of a clay source located some two miles from the Karekare town of Langawa. The clay is found at a spot where an intermittent stream forms a pool. During the dry season (when the pool dries out), women dig tunnels along the banks where rich and especially pure deposits of clay are located. The clay is loaded into large gourds and carried back to Langawa where (in 1971) old women process it for sale to neighboring artists.

Manufacturing clay into pigment is accomplished in seven stages: 1) raw clay is placed in a wood mortar and ground with a small amount of water; 2) the resulting mud is then transferred to another mortar containing water and a special bark; 3) the thick mixture is poured into a clay pot and allowed to stand until sand and other impurities settle; 4) the liquid is skimmed off and transferred to a large gourd bowl containing more bark and water;

5) the pigment that has settled at the bottom of this gourd is retrieved by skimming off the water with a small gourd dipper; 6) the remaining colored mud is spread out on a mat to dry; and 7) finally, the dried pigment is shaped into cakes for sale. This complex process yields a yellow pigment; to produce red, the ground raw clay of step 1 is rolled into small balls and burned before the procedure continues.

Artists buy these cakes of red and yellow pigment at the Langawa market where in 1971, they sold two for one penny. They produce their own black and white paints—black is made with ground charcoal or the soot left on the outside of a cooking pot; white was traditionally made from crushed animal bones (now commercial chalk is available in local markets). To produce more adhesive paints, artists mix these substances with sesame seed oil.

Oil-based paints are applied in three stages: the interior of the gourd is given an undercoating of red followed by one of yellow; second, by carefully scraping away areas of the yellow layer, a design emerges in red; and last, very fine black-and-white designs are painstakingly applied with a chicken feather (Pl. 20). The delicate white cross-hatched lines create the same intricate network of textures evident both in pyro- and pressure-engraving.

The settled Fulani also paint the insides of their pyro-engraved gourds (see Chappel 1977:38–49). Yet, in contrast to the dominance of this process in the Potiskum area (where some pressure- and pyro-engraving is done supplementarily as a result of Fulani influence), Chappel documented only one settled Fulani woman in the Yola area in the mid-1960s who practiced this technique. Thus, owing to their scarcity, gourds with designs painted on their interiors are especially prized by the Fulani.

Such Fulani drinking gourds, called *hasere*, are decorated with pigments applied with a stick or a corn stalk whose end is cut in the shape of a pen nib.[15] The colors used are the same as those found in the Potiskum area—black, red, white, and yellow. However, the sources of pigments, as well as their method of application, differ considerably:

> The inside of the gourd is first scrubbed clean and allowed to dry before the whole inner surface is painted black. The bark of a certain tree (*marehi*?) is crushed and put in a pot of cold water with the ashes of a type of grass (*lilaugi*?) used in the making of 'zana' mats. After a week or so the water begins to bubble up, a sign that the mixture is 'mature.' Ordinary wood ash is then added and the concoction applied to the interior of the calabash. Alternatively, the ashes of Indian hemp (*gabai*), mixed with oil from crushed melon seeds, may be used. It would appear that this black 'undercoat' serves as a primer, sealing the semi-porous inside surface of the gourd. When it is dry, the main areas of the design are painted in red. Cotton seeds are crushed to form an oily paste and mixed with *kadam* (red ochre), which can be purchased from the markets . . . Finally, the decorative motifs, including the 'framing' of the red areas, are painted in with white, or less frequently yellow, pigment. White is produced from crushed chalk or animal bones mixed with water or mahogany oil. Yellow is produced from *haire dala* (?yellow ochre) which is collected from the bush, crushed and mixed with water (Chappel 1977:38–40).

While such elaborate gourd painting is relatively rare, many groups uniformly coat the inner surface of drinking bowls with a thick layer of red or sometimes black pigment. If red ochre and oil are applied, the hardened surface can be polished with a pebble until it is almost like enamel. This renders the gourd watertight, makes it impervious to termite attack, and enhances its insulating properties (Chappel 1977:46).

Dyeing A number of groups favor gourds whose shells have been dyed or stained a different color, particularly red (Pl. 18). Otherwise unornamented calabashes are "rubbed with a concoction of millet leaves" to give them "a beautiful old rose color" (Trowell 1960:47). In contrast, others are polished frequently with indigo, achieving a very different color.

Ga'anda gourds are often considered unfinished until they are pressure-engraved and dyed red (Pl. 18d,e). The dye is produced by boiling the inner sheaths of guinea corn stalks (*hlarata*; *Sorghum caudatum var. colorans*) and indigo leaves (*kwechkwecheta*; *Lonchocarpus cyanescens*).[16] Calabashes are added while the mixture is still boiling, and they "cook" for a number of hours until the dye saturates. The dye is fixed by adding tamarind pods to the water. According to the Ga'anda, this process should only be done by prepubescent girls whose "pure" hearts allow a more beautiful red color to emerge (B. Rubin 1970:25). This method of dyeing accentuates the smooth untextured areas (*lifo*) and creates a striking contrast to the incised and blackened registers of design.

The settled Fulani also dye their gourds red after engraving the surface (Pl. 18a,f). Such gourds, called *bodere*, are colored using a similar compound made from the crushed leaves and outer sheaths of guinea stalks (*ngayori*) and the leaves of the indigo plant (*chachari*):

> Ideally, equal quantities of *ngayori* and indigo should be mixed, for too little indigo, which is said to act as a fixative, means that the dye will fade, while too much produces a dark red dye, verging on brown, which is held to be unpleasing (Chappel 1977:38).

The engraved gourd is immersed in this boiling compound for about one hour. Once removed, but before it has dried, a fixative made from wood ash slowly filtered through water is added. Bata gourds are colored in the same way, but instead of wood ash, the pods of the Egyptian *mimosa* are introduced into the boiling dye and produce a deep rich red (Pl. 18c).

While some Fulani gourds are color-impregnated in this manner, a resist-dyeing technique is most often employed so that certain areas of the surface remain uncolored. This produces the tessellated polychrome pattern characteristic of gourds made by the settled Fulani living around Garkida (where the Museum's gourds were collected) as well as in the Upper Benue region (Pl. 18a,f).[17] Most commonly, an oily paste made from crushed cotton seeds mixed with water is applied to the areas to remain undyed (Chappel 1977:38). The mixture adheres well and the oil prevents the dye from saturating the shell when the gourd is boiled according to the procedure described above. An alternate method was documented in Garkida where the areas to be left undyed were masked with a mud paste before a mixture of guinea corn stalks boiled with cottonseed oil was used to dye the gourd red (B. Rubin 1970:23).[18]

Gourds depicted in watercolor paintings by Carl Arriens and collected during the Frobenius expedition to the Adamawa region in the early twentieth century show the efflorescence of this combined pyrogravure/resist-dye technique (Pl. 19).[19] A range of striking examples was reproduced, some of which clearly represent the work of the settled Fulani who have dominated the Upper Benue region since the early nineteeth century.[20] Arriens claims the artist applied a paste made from "yellow organic powder" (ochre?) to areas of the unworked shell before immersing the gourd into a boiling solution of water and "the fruit of a poisonous type of millet" (1928:147). As a result of this process, the areas under the paste came out a glowing saffron yellow and the dyed shell a deep, bright red; both colors contrast with the black, engraved lines of design. It is noteworthy that only a few gourds in this large collection have painted interiors of the *hasere* category, suggesting that even at this early date the elaborate procedure was relatively rare. Other dyed gourds illustrated by Arriens, whose stylistic features are quite distinctive, were probably made by the Verre who live in the hills south of Yola.[21]

Additive materials A variety of extraneous materials can be added to the surface of any gourd. Its tractability allows patterns in wire or thread to be stitched directly onto the surface or holes to be pierced for various attachments. Gourds used for special occasions are often differentiated by such decorative additions. Although additive possibilities are not extensively exploited by peoples living in northeastern Nigeria, they do embellish dance rattles and some musical instruments. In other areas where gourds serve as prestige items in more stratified societies, they are often highly ornamented. This is especially true in the Cameroon Grasslands where attenuated bottle gourds are translated into royal attributes through the addition of elaborate beadwork.[22] Thus, an item of domestic utility is elevated into one of high social and political status through its beaded panoply of iconic designs.

DESIGNS

Although some women are known over wide areas for their skill, calabash decoration is not a specialization subject to hereditary or ritual constraints. In most groups, it is one of many talents a girl may acquire in preparation for her role as wife and mother (Chappel 1977:112). Even though gourd decoration is an activity any woman is free to learn, many prefer to commission others to do the work. Young girls interested in acquiring this skill begin at an early age, usually six or seven. Generally there is no period of formal apprenticeship; most begin by watching a close female relative at work. For this reason, it is customary to find daughters continuing the work of their mothers and grandmothers. Decorative techniques, particularly engraving and carving, are not mastered easily and usually take from two to four years to learn. As Chappel has observed, girls are not encouraged to begin practical training too early as the work involves a real element of physical danger—pyro-engraving requires a steady hand if burns are to be avoided, and pressure-engraving requires a high degree of control, as well as a good deal of strength to prevent cuts and gashes (1977:

64). Young apprentices generally begin by practicing on old or broken gourds, first drawing designs with charcoal (or earth) and water. Ga'anda students help their teachers execute the simpler, but time-consuming work of cross-hatching (*cerahla*) before they begin to engrave the lines (*njoxta*) that establish the gourd's design. They also help with other tasks such as blackening engraved lines and rubbing off the residue with dirt. Usually by their early teens, girls feel capable of executing a technique properly and producing satisfactory compositions.

Artists generally begin by decorating their own gourds and those of close relatives. Gradually, a woman may earn a reputation in the local community and subsequently receive commissions from others. Improved transportation networks and a weekly market system have broadened spheres of social interaction, thereby altering some traditional patterns of distribution and consumption. Artists rarely sell their decorative work at markets, but their skill can be observed in the impromptu exhibits that occur when women transport their goods to and from markets. Although some women have yielded to the prestige of modern factory-produced containers, many still use large splendidly decorated gourds as carrier-containers and for selling food in the traditional fashion. Thus, markets that draw women of many ethnic affiliations provide an opportunity to view different modes of calabash decoration (Pl. 6).[23]

Artists are often avidly interested in the work and dress of other groups, and heterogeneous ethnic contexts offer a wealth of inspiration. Chappel reports that one settled Fulani artist living in Yola incorporated a "Z" motif into her design vocabulary after seeing the pattern on a cotton gown in Girei market (1977:70). The Hausa men who frequently carve gourds for sale in northern markets—including the one at Yola—sometimes influence the work of local women. While the dynamic of intergroup contact provides a rich resource for decorative innovation, it makes the determination of ethnic provenance difficult.

The erosion of ethnic boundaries has brought neighboring groups into closer social contact. Additionally, in many areas there is a movement of divergent populations into large, burgeoning towns. A change of residence at marriage is also a common means of bringing together women of different backgrounds. All these situations have resulted in various kinds of technical and stylistic hybridization. For example, in the Ga'anda community of Boka, one artist (Dije) pressure-engraves lines so they re-create the broad cross-hatching of pyro-engraving, while not actually adopting the latter technique (Fig. 67c). Other Ga'anda artists occasionally transpose the compositions produced by pyro-engraving (such as those of the Bura; Fig. 67e) directly into pressure-engraving (Fig. 67d); gourds may also be dyed red, conforming to Ga'anda taste (B. Rubin 1970:25). In the Bura area—especially between the towns of Biu and Garkida—local women have borrowed the grids of settled Fulani gourd compositions (Fig. 67b; see Fig. 73). In turn, settled Fulani artists have copied Bura designs indicating that ideas flow in both directions (Fig. 67a).

Other technical adaptations are evident in the area around the Gongola-Hawal confluence, where Dera, Gbinna, and Bura populations live in close proximity. A Gbinna woman may maintain her own technique of pressure-engraving but afterwards have a Dera woman abrade the surface of a decorated gourd with a soft stone (Fig. 68c). Other Gbinna women have entirely adopted the Dera mode and style of pyro-engraving (Fig. 68b,d). Likewise, some Dera artists have borrowed the engraving style of Bura women who now populate Dera towns in large numbers (Fig. 68a).

In the Potiskum Plains, it is common for local women to buy gourds pressure-engraved on the outside by the pastoral Fulani and then to finish the interiors by applying pigments in the traditional fashion. A number of Karekare, Ngizim, and Ngamo women have even learned pressure-engraving and thus can produce their own Fulani-like designs on the exterior (Figs. 57,58).

With these dynamics of production and distribution as background, some of the aesthetic choices that determine design systems will be presented below. While conventions exist for constructing decorative compositions, considerable stylistic interpretation within groups accounts for the kaleidoscopic variety of gourd designs. Due to the nature of the Museum's collection, the stylistic variations emphasized here will be those developed by peoples using the two dominant technical processes found in the northeast—pressure- and pyro-engraving. The following analysis must be prefaced with the admission that generalizations cannot possibly embrace all the decorative nuances that distinguish this remarkably diverse art form.

There are two obvious formal factors that have influenced the development of design compositions: the curve of the gourd surface and the technique(s) used to decorate it. First of all, whether a gourd is a bowl, bottle, spoon, or tube determines how a

67. Decorative hybridization. Clockwise from top left.
a. Fulani pyro-engraving in a "Bura" style. 18.1 cm x 18.7 cm. UCLA MCH X83–742;
b. Bura pyro-engraving in a "Fulani" style. 21.3 cm x 21.3 cm. UCLA MCH X83–683;
c. Ga'anda pressure-engraving in a "pyro-engraved" style (*yaraxwata*). 20 cm x 8.9 cm. UCLA MCH X85–24;
d. Ga'anda pressure-engraving in a "Bura" pyro-engraved style. 15.2 cm x 18.2 cm. UCLA MCH X83–734;
e. Bura pyro-engraving style. 21.6 cm x 23 cm. UCLA MCH X83–737.

68. Decorative hybridization.
a. Gasi (Dera subgroup) gourd pyro-engraved in a "Bura" style. 23.5 cm x 25.1 cm. UCLA MCH X83–738;
b. Gbinna gourd pyro-engraved in a "Dera" style. 17.6 cm x 19.7 cm. UCLA MCH X85–37;
c. Gbinna pressure-engraved gourd with Dera "stone-washing." 21.3 cm x 22.5 cm. UCLA MCH X83–656;
d. Gbinna gourd pyro-engraved in a "Dera" style. 12.7 cm x 13.7 cm. UCLA MCH X85–38.

design will be oriented on its surface. With hemispherical bowls, the uniform shape places a natural emphasis on the center. Most groups use this structural constant as an organizing principle, with the rest of the composition symmetrically arranged around it. When the center of the gourd is the design focus, it is often emphasized by a series of circular registers or radially aligned shapes (Pl. 22). The Ga'anda, who seem to prefer ovoid-shaped bowls to spherical ones, tend to coordinate the ornamentation with the relative length/width ratio of individual examples (Pl. 23). Ga'anda artists consistently remarked that even if clients requested specific designs, each gourd's shape was the key factor in determining a compositional arrangement. Most groups use the gourd's round or ovoid shape as a guide for dividing the design field into a number of symmetrical units, but some prefer an equally harmonious format that treats the surface as one continuous design field (Fig. 4). The pastoral Fulani, who generally decorate deeper gourd bowls, concentrate their designs around the broad side panels where motifs are arranged to converge, like radii, on an unmarked center (Fig. 69; Chappel 1977:44).

This vertical design format suggests that, for some groups, the alignment of motifs is adjusted to conform to how the gourd is to be viewed. For example, placing the designs around the rim of a Fulani milk container affords maximum visibility when the gourd is carried upturned on its base (Fig. 12). This compositional focus seems to be as much a response to the importance of display as to function. The same may also be true of certain large Tera gourds which, unlike those decorated with an emphasis on the center, have designs arranged around the margins in "Fulani" fashion (Pl. 24). This compositional variation may have developed in response to the way they are displayed in women's rooms—nested, with only the edges prominently visible (Pl. 9). Gourd spoons exhibited in Tera rooms are also in full view and thus are usually decorated along their entire lengths with designs appropriate to the shape of their cups and narrow handles.

The second factor governing the evolution of design systems is the constraints imposed by each of the two decorative techniques emphasized here, pressure- and pyro-engraving. The iron tool used in pressure-engraving is gripped like a dagger and pulled across the surface with the wrist locked in position, producing only relatively short lines across the gourd surface (Pl. 17; Fig. 53). This technique encourages compositional formats based on small units of design filled with fine linear patterns (Pl. 23). For example, the overall patterning used by the Longuda is consistent with the simple, tight, and controlled movements used in executing the design (Fig. 70). On the other hand, the mechanics of pressure-engraving discourage certain kinds of designs, such as those that emphasize continuous lines (Fig. 71). While the effects possible through pressure-engraving are technically limited, there is still ample freedom to make motifs that are simple or complex, regular or irregular.

Pyro-engraving, on the other hand, provides the opportunity to work with broad, sweeping gestures (Pl. 16). The size and shape of the knife blade and the need to work rapidly while it is hot mean that precise, rhythmic movements are best suited to this technique (Chappel 1977:34). As is true for pressure-engraving, the position of the hands dictates how designs are executed. Chappel's general assessment of this technique is worth citing here:

> The most satisfactory solution seems to be to establish a formal, geometric framework by cutting lines parallel to the diameter and stretching across the central area of the inverted gourd which remains firmly in the carver's field of vision, and which can be easily covered with single, rapid movements of the knife . . . As the design field is divided and sub-divided a framework is established, each geometric shape generating other geometric shapes until the area is filled in, by a process of reduction, with typical geometric forms which have, as it were, developed out of one another (1977:34).

This characterizes the engraving process used by most of the groups represented in this collection, illustrated here with a Dera example (Fig. 72). However, Chappel's description closely describes the rather formal, geometric grids drawn and filled by the settled Fulani, about whom he is writing (Fig. 73). It also neatly explains the way many modern market gourds are executed (Fig. 74).

Within the decorative parameters imposed by each technique, diversification is achieved through the spatial distribution of design motifs and the interplay between them and the unworked areas of the gourd's surface. Regardless of technique, the general tendency is to treat the gourd surface as a single compositional field delimited by a frame parallel to the rim edge.

69. Drawing of pastoral Fulani gourd (see Fig. 25).

70.
Longuda gourd bowls (*kwarawa*).
a. 21.9 cm x 23 cm.
UCLA MCH X83–657;
b. 19 cm x 20.9 cm.
UCLA MCH X83–660;
c. 19.7 cm x 20.4 cm.
UCLA MCH X83–659;
d. 22.2 cm x 22.5 cm.
UCLA MCH X83–658.

71. Ga'anda pressure-engraved gourd bottle (*buta*).
34.5 cm. Berns collection.

72. Dera woman filling in sections of a pyro-engraved composition; Buma. December 1980.

73. Pyro-engraved "grid" compositions of the settled Fulani (*bodere*).
a. 26 cm x 27.3 cm. UCLA MCH X83–762;
b. 24.1 cm x 25.1 cm. UCLA MCH X83–789.

74. Bura market gourds purchased in Biu and Wandali, 1970.
Clockwise from top center:
a. 11.4 cm x 12.7 cm. UCLA MCH X83–747;
b. 13 cm x 14.4 cm. UCLA MCH X83–750;
c. 13 cm x 14.3 cm. UCLA MCH X83–759;
d. 10.8 cm x 11.4 cm. UCLA MCH X83–752;
e. 12.4 cm x 13.3 cm. UCLA MCH X83–756;
f. 9.4 cm x 9.7 cm. UCLA MCH X83–758;
g. 12.5 cm x 13.5 cm. UCLA MCH X83–751;
h. 13 cm x 14 cm. UCLA MCH X83–755;
Center gourd: 20 cm x 20.6 cm. UCLA MCH X83–739.

Pyro-engraving As indicated above, pyro-engraving allows the artist to block out areas of design and then progressively fill them in with a range of linear formations (cf. Fig. 72). While the same tendency to divide the design field into a series of geometric registers predominates in pressure-engraving, there is a basic figure/ground reversal that causes the incised "blackened" areas to appear as background and the light strips of unworked gourd as relief designs. The settled Fulani often use this compositional device to establish a grid across the design field, which is defined by strips of unworked shell that remain after the principal decorative areas have been blackened (Fig. 73).[24] This formal organization of motifs results in minimum tension between figure and ground. Instead, the dynamism of settled Fulani designs derives from the range of motifs used to fill the grid framework and from the introduction of resist-dye patterns (Pl. 18a).

Formal linear grids established by strips of unworked shell also dominate design compositions of the Kilba (see Fig. 48). However, rather than using a variety of motifs to fill sections into which the design field is divided, they tend to rely on a variety of crosshatched textures. Gourds decorated by the Chibak also are characterized by fine linear grids; compositions are invigorated by the bold shapes that enclose them and their juxtaposition against broader areas of unworked surface (Pl. 28a). The example illustrated here has four symmetrical geometric units whose curvilinear sides provide a kind of template for the five large circles positioned between them. These discrete shapes are further integrated by deeply engraved lines framing the main design units, a device that works in much the same way as do concentric hatched bands in pressure-engraving. Compositional balance is also achieved by the repetition of various finely hatched textures, such as the one filling the five prominent circles, which is intriguingly similar to the rattan caning used on the seats of Western chairs.

The division of the gourd's surface into a number of symmetrical design units also is characteristic of Dera compositions. However, the arrangement of geometric shapes is more flexible than that of the preceding groups, and the amount of surface area covered with engraved lines more wide-ranging. The gourds collected and documented among the Dera, as well as the neighboring Pidlimndi, fall into two main stylistic categories.[25] The first conforms quite closely to gourds described above—the surface is divided into a number of units, dominated by large circles filled with densely engraved patterns (Pl. 22; Fig. 75c). The design composition emerges from the circumscribed strips of unworked relief. In the second style of Dera/Pidlimndi engraving (Pl. 26; Fig. 75d), a number of bold curvilinear shapes are burned in, their tangential angles creating the same circular framework that dominates the first style. Here, however, the drawn shapes stand out more dramatically, their contrast with the gourd shell intensified by its being abraded (i.e., made lighter) with a soft stone. While the two styles of engraving used by the Dera may be quite different, it is apparent that what separates them is an interesting spatial inversion—the abraded shapes that dominate as figures in one are filled with dense lines to become the background in the other. The juxtaposition of circles with three- or four-sided curvilinear shapes is another diagnostic feature of many Dera gourds. Their syntactical balance is nowhere more harmonious than in the all-over patterning of Plate 27 and Figure 75a,b where the interlocking motifs can achieve a lively flip-flop between figure and ground.

Unlike the above compositional arrangements, those of the Bura, Tera, and Waja are divided into central and marginal decorative fields. The design focus is defined by a central circle framed by concentric or vertical registers filled with symmetrical patterns (Fig. 76). The same horizontal and vertical banding characterizes motifs engraved around bottle-shaped gourds decorated by these same groups (Pl. 21; Fig. 14). Whereas the most striking design element in pyro-engraved gourds is usually the contrast of the worked/unworked areas, this impact is reduced in these examples by the relatively even coverage of the surface and the rather modest interplay between figure and ground. The Bura use complex circular patterns as the focus of their densely engraved compositions (Pl. 25), an approach quite different from that of some Tera artists who use a multiplicity of fine lines (*ndesa*) in a dynamic arrangement to increase the decorative effect of the basic organizational format (Pl. 29; Figs. 77,78). Furthermore, the Tera, Jera, and Waja engrave gourds with a vertical rather than concentric orientation of motifs (Pl. 30). These spokelike units, each of which represent a figure drawn against an unmarked ground, are similar to those used by some pastoral Fulani in pressure-engraved compositions (cf. Fig. 69).[26] Yet, this derivative style reveals its debt to the circular format by sometimes having a series of engraved (as well as carved) bands framing the rim edge (Pl.

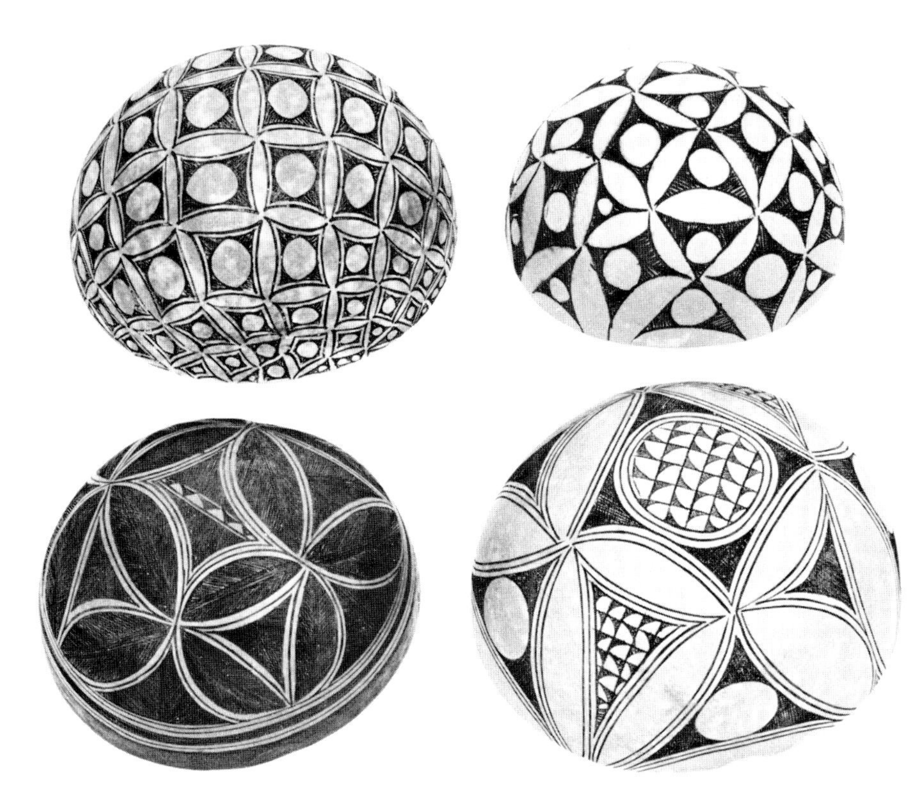

75. Pyro-engraved gourds.
a. Pidlimndi. 20.9 cm x 22.9 cm.
UCLA MCH X83–676;
b. Dera. 18.1 cm x 20 cm.
UCLA MCH X85–33;
c. Dera. 19.4 cm x 20.6 cm.
UCLA MCH X83–680;
d. Dera. 20.9 cm x 23.5 cm.
UCLA MCH X83–677.

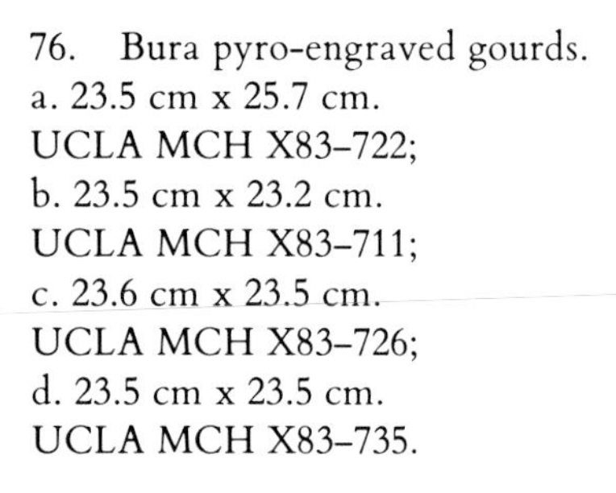

76. Bura pyro-engraved gourds.
a. 23.5 cm x 25.7 cm.
UCLA MCH X83–722;
b. 23.5 cm x 23.2 cm.
UCLA MCH X83–711;
c. 23.6 cm x 23.5 cm.
UCLA MCH X83–726;
d. 23.5 cm x 23.5 cm.
UCLA MCH X83–735.

77. Tera pyro-engraved gourds with *ndesa*, all decorated by the artist Jumai Pitiri Gulcoss; Wuyo.
a. 23.5 cm x 22.8 cm. UCLA MCH X83–730;
b. 24.1 cm x 24.7 cm. UCLA MCH X83–706;
c. 24.1 cm x 23.5 cm. UCLA MCH X83–687.

78. Tera pyro-engraved gourds.
a. 7.6 cm x 8.4 cm. UCLA MCH X83–696;
b. 19 cm x 20 cm. UCLA MCH X83–728;
c. 14.9 cm x 16.2 cm. UCLA MCH X83–714;
d. 6.3 cm x 7.5 cm. UCLA MCH X83–697.

79. Mbula pyro-engraved gourds (*gilu kwar*).
a. 22.5 cm x 23.5 cm. UCLA MCH X85–45;
b. 22 cm x 23.5 cm. UCLA MCH X85–47;
c. 20.3 cm x 21.6 cm. UCLA MCH X85–43;
d. 24.5 cm x 24.5 cm. UCLA MCH X85–44.

30). In other Tera examples, instead of a center virtually free of design, an inner circle of radial motifs achieves the same dense two-part coverage of the surface as in the conventional concentric format (Pl. 29; Fig. 77c). It is evident that even the range of compositional diversity seen in Tera, Bura, Jera, and Waja gourds conforms to certain basic design principles.[27]

The Mbula, singled out earlier for their interesting technical interpretations, are also notable for their design compositions. Each Mbula gourd is dominated by an unusual asymmetrical arrangement of bold, adjoining geometric shapes (Fig. 79). Compositional balance is achieved by aligning at least one set of identical motifs across a radial axis. The outstanding feature of Mbula gourds is the strong reverse relief of the heavily engraved and scorched design elements. The careful outlining of the reserved spaces and the addition of spiral links seem to minimize tension between figure and ground. Even the small sketch of a single scorpion in Figure 79c is placed to maintain a balanced design.

Pressure-engraving In pressure-engraving, the figure/ground relationship is established by incising designs against the gourd shell. The spatial relationship between the drawn elements, however, varies considerably depending on the conventions of different groups. The engraved work of the pastoral Fulani, for example, takes two forms.[28] The first, described above, involves a vertical arrangement of motifs around the outer area of the gourd leaving the center free (Fig. 69). The composition is organized in four sections, each dominated by a larger design element that reaches toward the center and is situated at the termini of the gourd's horizontal and vertical axes. The second format also involves dividing the design field into four parts comprised of primary and secondary motifs, but here greater use is made of the gourd's surface, as well as the technique of broad-groove engraving (Fig. 80). The relationship between worked and unworked areas is quite different in the two pastoral Fulani styles, but the quadrapartite divisions common to both provide the same compositional balance.

80. Drawing of pastoral Fulani pressure-engraved gourd (*tuppande*).
24.1 cm x 27.9 cm. UCLA MCH X83–782.

The Hona, Ga'anda, and Gbinna generally divide pressure-engraved compositions into four symmetrical units. Each group, however, resolves the spatial relationships of motifs quite differently. Hona examples reveal the most dynamic contrasts between worked and unworked areas (Pl. 31). While the radial organization of motifs may be similar to that used by the Fulani, the vigor injected into this otherwise formal arrangement comes from the combination of shapes with curved and straight edges. This spatial alternation allows the bold reserved forms between them to project as "figures" against the engraved ground. The interplay is further enhanced by the tight geometric patterns filling the "drawn" shapes. This same striking figure/ground relationship is found even on small gourd spoons (Fig. 13c,d).

Compositions on Ga'anda gourds are also based on the symmetrical alignment of motifs, but their spatial arrangement contributes to a more static design. For example, the Ga'anda calabashes in Plate 23 incorporate the same curved motifs that dominate in the Hona gourds described above, but by aligning the spatial units beside them to echo the same contours, the result is less distinct secondary patterns. Additionally, the use of linear banding works to stabilize the composition, as does the rigid patterning of the filler motifs, both absent in most Hona gourds. While Hona artists have interpreted the conventional designs and filler patterns they share with the Ga'anda more freely, the Ga'anda range of compositional solutions is far more extensive than those found among the Hona.[29]

The elaborate compositions on Gbinna pressure-engraved gourds occupy an intermediary position between the vigorous arrangements of the Hona and the formal patterns of the Ga'anda (Fig. 81). Figure 81a shows clearly how the bold contrast between worked and unworked shapes of the central design area (related to that of Hona gourds, but here intensified by the texture resulting from Gbinna deep gouging) is held in check by alternating bands framing the inscribed motifs (a convention also used by the Ga'anda). In sum, Ga'anda, Hona, and Gbinna designs rely on the complex interplay between figure and ground, and all exploit the range of decorative patterning and textural variation that can be achieved through pressure-engraving.

Longuda gourds perhaps best illustrate the complementarity between design and the process of pressure-engraving (Fig. 70). Most are densely engraved with minute oval designs that envelop the gourd surface like a delicate mesh. The strict balance between the pattern and its engraved ground allows little of the formal interplay seen in gourds decorated by the Hona, Ga'anda, or Gbinna. Instead, the highly repetitious, precisely executed designs are animated by the sheer complexity of their interlace patterning (Fig. 82).

MEANINGS

Among groups living in northeastern Nigeria, compound designs covering the surface of a gourd are often given names corresponding to distinctive decorative patterns. Whereas individual meanings have accrued to particular units of design, their combination rarely adds up to a coherently meaningful whole. Motifs usually do not relate to one another in any thematically consistent way. Instead, heterogeneous names are given to highly abstract designs that fundamentally serve as visual equivalents of a wide range of human experience.

In attempting to elicit names for composite motifs, it became apparent that most observers and many artists are familiar only with the most conventional and distinctive designs. Often a name given to an elaborate composition by an informant is drawn from a dominant motif. The degree of equivocation in explaining design compositions suggests that verbal associations do not govern how motifs are combined nor how the finished product is evaluated. For example, in the course of Chappel's extensive research, he concluded that

> a gourd design is perceived, and conceived, in the first instance, as a composite arrangement of linear patterns undifferentiated as to its component elements; for it was apparent that observers first reacted towards the visual stimulus of the overall structure of the design and then only, if at all, did they differentiate the pattern elements for the purpose of identification and interpretation in verbal terms.
>
> Local observers, it seems, do not expect to find a clear-cut subject-content in gourd designs; therefore they do not generally look for one. An overt interest in the subject-matter of designs and pattern elements was only apparent when I myself tried to elicit meanings from observers. Even then the level of interest which a request for meanings aroused tended to vary from individual to individual and from area to area (1977:54).

It seems likely that scholars are the ones most interested in deciphering gourd designs so that such decoded messages can be used to better understand cultural processes. And for good reason, since

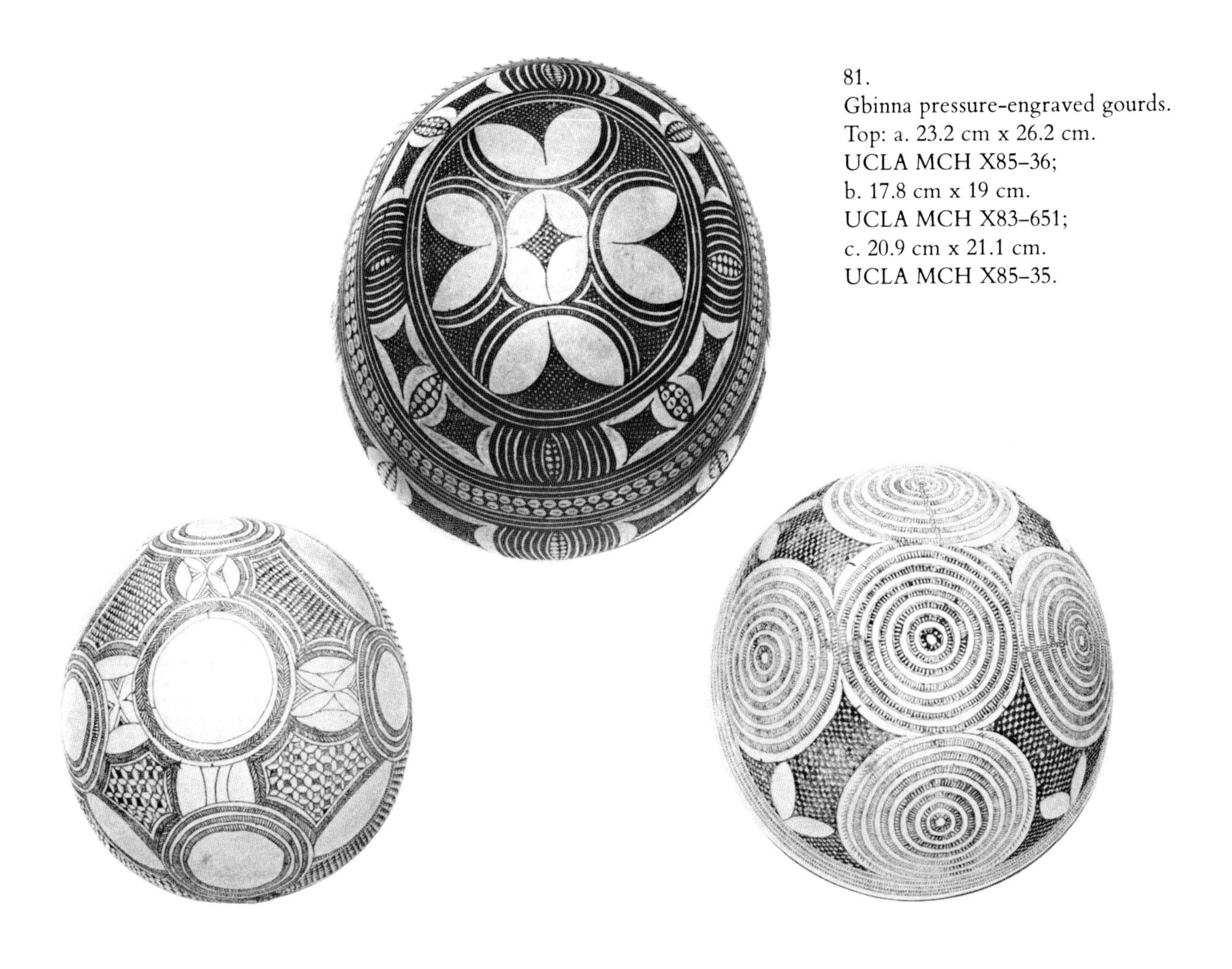

81.
Gbinna pressure-engraved gourds.
Top: a. 23.2 cm x 26.2 cm.
UCLA MCH X85–36;
b. 17.8 cm x 19 cm.
UCLA MCH X83–651;
c. 20.9 cm x 21.1 cm.
UCLA MCH X85–35.

82. Longuda pressure-engraved gourd (*kwarawa*); Guyuk. October 1981.

83. Waja pyro-engraved gourds.
a. 22.4 cm x 23.8 cm. UCLA MCH X85–52;
b. 23.2 cm x 22.7 cm. UCLA MCH X85–48;
c. 17.8 cm x 20.3 cm. UCLA MCH X85–50.

it is evident that the names *consistently* given by a group for conventional design patterns *do* provide important insights into the kind of visual links made between forms and their referents, even if artists do not intend to introduce "meaningful" subject matter. Therefore, the shapes and designs produced should not be construed as only pleasing arrangements that satisfy conventional aesthetic standards, even though it is difficult to determine and verify the meanings of designs.

Another issue is whether form precedes content, and while a legitimate query, it diverts attention from the task of understanding the relationship between human experience and decorative shapes.[30] Even if it is taken that there is no single correct name for an individual pattern element, it is possible to evaluate the kinds of associations suggested by the designs. Verbal meanings attached to particular decorative patterns often alter with shifting historical, social, or geographical circumstances indicating that designs as well as their referents are dynamic, rather than fixed, variables. While only a few general comments and a limited number of specific examples will be offered to elucidate the complex subject of gourd iconography and its implications, it is clear that the graphic language used in decoration has developed out of larger cultural processes. The chapter on the Ga'anda will show how gourd designs and the meanings assigned to them are best understood within the ethnographic framework in which they have evolved.

Traditional referential iconography used in gourd decoration can be divided into two general categories—designs that relate to the natural world and designs that relate to the cultural environment. In both, names are used to differentiate drawn shapes, the reserved spaces left between them, and the patterned motifs that fill them. In the first iconographic category, names like "guinea corn," "honeycomb," "turtle's back," "lizard's claws," and "guinea fowl feathers" are used to identify shapes or patterns that correspond to core visual characteristics of their natural referents. There are notable variations, however, in the organic equivalents that groups associate with particular designs. For example, "guinea corn" is given by the Mbula to the zones of cross-hatching (*mesa*) commonly used in their compositions (Fig. 79). The same referent is used by the Waja for describing strips of complex hatching (*gomo jemma*) that encircle the outer rim of their gourds (Fig. 83a). Yet, when using this motif to decorate the center of the design field, the Waja call the pattern of hatching "honeycomb" (*gemen yaro*; Fig. 83b,c). This iconographic distinction results from the striking differences in alignment, number, and gauge of the hatched lines or from the shape of motifs that enclose them. Similarly, the Gbinna incorporate densely hatched patterns in their gourd decoration, which in pressure-engraving result in textured rather than mere linear grids (Fig. 81). They, however, identify this black-and-white checkerboard motif as "guinea fowl feathers" (*tu'a-tu'a*), as do the Yungur (Chappel 1977:figs. 84–86).

The multiple referents offered for this common decorative device illustrate the number of visual associations groups make with similarly patterned elements in the natural world. Cross-hatching is a technical procedure, but the pattern it creates presents a stylized map of the distinctive features that distinguish particular natural forms. When called "guinea corn," it describes the cluster of corn seed at the stalk's head. The design correlation between cross-hatching and a honeycomb grid is even more obvious. And, the association between "guinea fowl feathers" and the relief effect of Gbinna and Yungur pressure-engraving seems to be most strongly based on the feathers' patterned color contrast.

The second category of iconographic referents identifies how man has altered nature to suit his own cultural needs. Such design names refer to styles of coiffure, patterns of body and facial scarification, modes of ornamentation, and other technical processes. Here, too, links are established by means of the distinctive shapes and patterns shared by designs and their referents. For example, as suggested above, certain motifs in pastoral Fulani gourds are depictions of facial scarification.[31] Likewise, radial motifs on Tera and Waja gourds are identified with these peoples' complex programs of facial striations (Pl. 30f; Figs. 77b,83). Thus, this second category of referential iconography can distinguish gourds of different ethnic groups because designs are often specifically associated with distinctive features of self-decoration. Additionally, the sawtooth pattern frequently framing the rim or center of a gourd is named "chipped teeth" by the Yungur, who alter their upper and lower incisors in this fashion.[32] Accordingly, their Ga'anda neighbors identify this same decorative motif on their own gourds as *hlan'yungurata* ("Yungur teeth" or "Gbinna teeth") because they do not maintain this tradition (see Fig. 98). Artists in the Tera villages of Kwadon and Kwaya Kusar call this motif *zhinda tangala* ("Tangale teeth") to ridicule this Tangale practice. The Hona, on the other hand, call these recurrent triangles *hlan'pishara*, "teeth of a hyrax," using a zoomorphic referent (see Pl. 31d).

Gbinna gourds identify other distinctive cultural features: the central reserved shapes circumscribed by lines and framed by patterns of "guinea fowl feathers" are named after a particular style of women's hair shaving, called *ton'ton'ja* (Fig. 81a). The hatched circular bands of design that divide the composition into registers are "plaited ropes" (*kurta*), and the small repeated bisected ovals aligned between them are iron ornaments (*gwaro*). Yet, the meaning ascribed to a decorative pattern does not necessarily imply that it literally describes the object in question; rather, it captures some essential visual aspect of it. For example, the central patterns of these Gbinna gourds (Fig. 81a) do not exactly reproduce the coiffure known as *ton'ton'ja*; instead, the circular arrangement of identical shapes is a stylized equivalent of the hair tufts that distinguish it. Likewise, the "ropes" of design used to divide registers of the Gbinna gourd take their name from the linear texture common to both plaiting and cross-hatching. That designs suggest their referents rather than literally reproduce them is supported by the fact that a very different Gbinna design composition, five sets of concentric circles, is also named *ton'ton'ja* (Fig. 81b).

A gourd decorated by the Ga'anda (Pl. 18e) has a central band dividing the composition, called *temtedera*. This name refers to a children's coiffure where the hair is shaved off except for one strip left along the top of the head. The four circles engraved on either side of this band are called *pwoti*, the Ga'anda inverse of the Gbinna *ton'ton'ja*, where four circles of hair are shaved off the top of a woman's head. The combination of two coiffure patterns—one exclusive to children and the other to women—on a single decorative field, indicates that design choices are not governed or evaluated by how "meaningfully" they reproduce their iconographic referents. Moreover, while both designs aptly capture the correct contours and alignment of their visual equivalents, the checkerboard patterns they circumscribe (*wanfiyca*) are equally important to the design composition. Unlike the Gbinna who identify this motif with "guinea fowl feathers," for the Ga'anda this pattern is a graphic transposition of the unusual variegated weave of small cloth aprons (*wanfiyca*) worn by priests.

These examples show the composite character of gourd iconography and suggest that most named referents bear little contextual relationship to one another. They do, however, consistently infer a sensitive response to the visual properties of things. Groups vary in the number of distinctive patterns they differentiate by name. Additionally, the number of motifs a group incorporates ranges widely, with those used by the Ga'anda among the most numerous and complex. Where designs are elaborate, often involving compound meanings such as those referring to shape and those referring to internal patterning, there tends to be more variation in interpretation. On the other hand, Longuda gourds, dominated by a single stylized motif, are consistently identified in the same way—cowrie shells (*dumbela*; Figs. 70,82).

While the meanings of certain conventional patterns, possibly representing older designs, have remained fairly constant, others have changed with time. For instance, Hudson observed that among the Tera, components of a traditional referential iconography have given way to a more technical vocabulary describing the kinds of lines engraved rather than what they mean. The Dera, Bura, and Waja also name particular design elements according to the way lines are engraved. All these groups differentiate between the thick black lines achieved with the broad edge of the blade and the tiny dots produced with the tip (Fig. 83a,b). Bura and Dera artists identify cross-hatching as *tankrinkta* and *dadaula*, respectively, and no longer attach any organic reference. Likewise, the multiplicity of fine lines characteristic of Tera engraving are named *ndesa* ("feathery lines"); this word seems to bear little relationship to designs formed by means of this decorative device (Pl. 29; Figs. 77,78). But despite the emergence of a technical vocabulary, it is unclear whether meaningful designs have completely "given way" to these procedural referents. It may be that multiple associations have accrued over time to shapes as well as to the lines that enclose and fill them.

It should be reiterated that the interest of artists in the work of others, from the same or another ethnic group, has resulted in the introduction and reinterpretation of motifs and techniques. The names offered for gourd designs help trace sources for particular derivative motifs. For example the Tera may transpose a design into *ndesa* and give it a name defined as "the design that Ladi copied from the Z people" (B. Rubin 1970:22). This phenomenon seems to be prevalent in areas where different, but related, groups live in close proximity. In the Uba Plains, the Nggwahyi, a group related to the Bura, exhibit a style of decoration that is a composite of ideas borrowed from their Kilba, Chibak, and Bura neighbors. To differentiate decorative patterns, they are given specific names such as "copied from the western people," which indicates a Bura source. Among the eastern Margi, women consciously imitate designs of others and one popular geometric pattern is named accordingly, "Fulani" (Vaughan 1975:185).

The meanings of certain decorative patterns have changed in response to historical and social circumstances. The "dots and dashes" frequently used by the Waja are associated today with the vowel markings of Arabic script (*jayya* and *jokku*; Fig. 83a,b). This identification likely resulted from the increasing Islamization of the Gongola-Hawal confluence and the consequent availability in local markets of the Koran written in Arabic. Another example of how a referent assigned to a conventional gourd design is altered to reflect changing circumstances involves a series of juxtaposed chevrons used by various groups. The Ga'anda, for example, give it the name "fish bones," a traditional organic referent (see Fig. 94h). One male Yungur carver identified the same motif as "sergeant's stripes" (Chappel 1977:54; figs. 100–101). In the latter instance, perhaps verbally translating this common motif into an exclusively masculine emblem of social status helps legitimize Yungur men's participation in a traditionally female craft.

Translations of traditional motifs into modern visual equivalents also reflect the impact of imported commodities. For example, dense hatching that leaves only small dots of shell visible is now identified with the pattern on the bottom of plastic sandals or "zoris," as well as their distinctive footprints (Fig. 84; see Chappel 1977:fig. 99). Small circular spirals with a number of organic associations (e.g., fish eyes or leopard's eyes) have been identified alternatively as wheels of an automobile or "headlights of the District Head's motor car" (Fig. 85; Chappel 1977:figs. 93,96). The prestige of car ownership in modern Nigeria has led the settled Fulani to rename a common traditional motif, "motor car"—a square with spiral "wheels" at each corner.

In addition to attaching new meanings to old forms, changing motifs reflect new ideas and modern technology. For example, the painted interior of one gourd made by the Karekare juxtaposes Koranic prayer boards and what were identified as helicopters, but look more like airplanes (Pl. 32). The use of a Moslem emblem invoking the presence of Allah can be interpreted as an acknowledgment of changing religious affiliations.[33] At the same time, the "helicopter" motif may symbolize not only Nigerian modernization but also political aspiration, since it has been used on gourds

84. Yungur pyro-engraved gourd (*kũsa*). 17.8 cm x 20.3 cm. UCLA MCH X85–27.

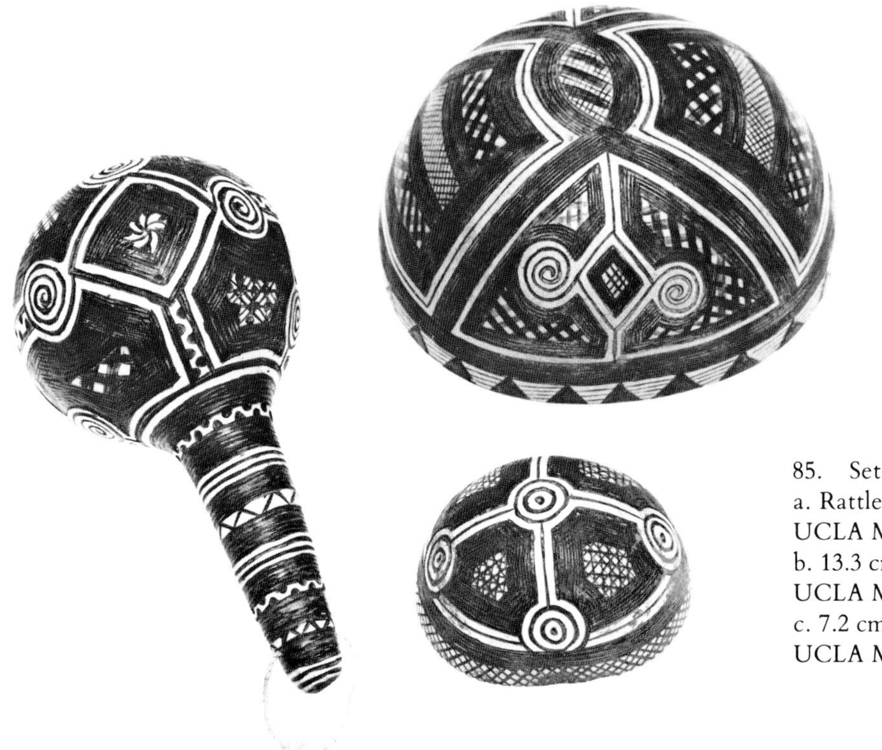

85. Settled Fulani "headlight" motif.
a. Rattle. L. 16.5 cm.
UCLA MCH X83–748;
b. 13.3 cm x 15.1 cm.
UCLA MCH X83–765;
c. 7.2 cm x 8.5 cm.
UCLA MCH X83–678.

86. Kanuri pyro-engraved gourd
with British soldiers and a camel caravan.
35.5 cm x 36.4 cm. UCLA MCH X85–56.

made in the Potiskum area since 1959, when Chief Obafemi Awolowo, campaigning for his "Action Group," flew over the area trailing a banner with the party's initials on it. Such graphic references to recent social changes, each inspired by a quite different source, are combined with traditional patterns, revealing how gourd decoration mirrors the syncretism of modern life. The helicopter depicted on the Karekare gourd, for example, is built up of designs that can be identified as "guinea fowl feathers," "mat weave," and "tortoise shell."

Although gourd designs tend to be abstract equivalents of the visual world, the last example introduces a category of more representational forms. Among most groups, such motifs often are exceptional and appear rather randomly: e.g., a scorpion drawn on an Mbula gourd (Fig. 79c), a lizard on a Waja bottle (Fig. 14d), or human figures on a pastoral Fulani henna tube (Fig. 17) Hausa men from the Kano area pyro-engrave gourds for sale in the larger northern markets and draw from a large repertoire of animal motifs in their decoration (see Fig. 151).[34] Yet, of the gourds that incorporate such representational imagery, those made by the Kanuri living in the Maiduguri area are the most intentionally narrative (Figs. 86,87). They include heterogeneous scenes with British soldiers, camel caravans, and royal equestrians depicting in literal, albeit stylized, language the phenomena that have characterized the history of this geographically important region south of Lake Chad.[35] Certain decorative devices are highly typical of the Kanuri and may make references to visual equivalents in other media. For example, stylized camels and horses with their long curved necks may refer to the clay animals with small mounted wax figurines that are commonly made as children's toys by groups in this area (see Krieger 1969 II:210–211,214). The stylized British soldiers seem to have been copied from caricatures of palace guards imported from England, rather than from the soldiers themselves. Likewise, Figure 88, a highly abstract Kanuri gourd design, imitates women's elaborate coiffures (Fig. 89) that are also reproduced on wax figurines or dolls commonly made by groups living in northern Nigeria (see Krieger 1969 II:figs. 212–213; Chappel 1977:fig. 216 l,m,n).

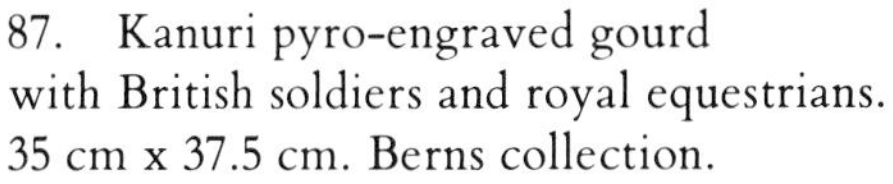

87. Kanuri pyro-engraved gourd with British soldiers and royal equestrians. 35 cm x 37.5 cm. Berns collection.

88. Detail of Kanuri gourd (see Pl. 28).
28.7 cm x 29.5 cm. UCLA MCH X83–797.

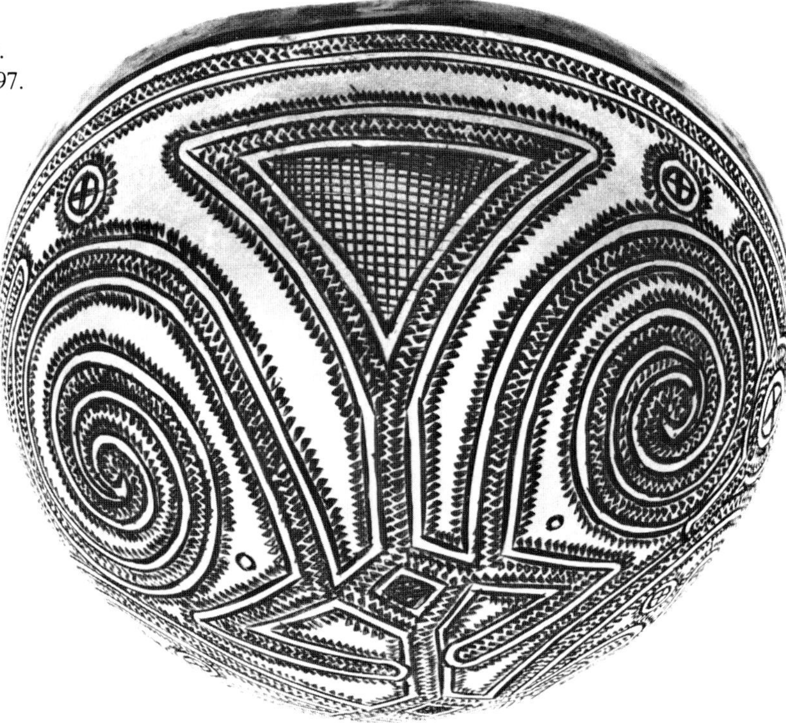

Although representational imagery is relatively rare in the work of northeastern Nigerian artists, three unusual gourds in the Museum's collection illustrate the narrative possibilities in gourd decoration (Pl. 33; Figs. 90–92). According to accession records, they were displayed at the 1924 British Empire Exhibition at Wembley. The label affixed to Figure 92 identifies the artist as a Hausa/Fulani man (Audu Mai Alijeta) from the town of Jalingo, located southwest of Zinna in Gongola state (see Map B). The stylistic correspondences between the three gourds indicate that they all were the work of the same artist.

The figures that dominate the compositions of these gourds are splendidly diverse. They range from unclothed "natives," to elaborately robed Hausa musicians (depicted with a strikingly Eastern flavor), to precisely uniformed colonial soldiers, primly dressed Christian missionaries, and a carefully posed ballerina. Also distinctive are the motor cars in Figure 91b, the larger of which appears to be based on a British Austin 7 dating from the mid-1920s.[36]

While the keen sensitivity to detail and texture evident on these gourds is typical of other northeastern Nigerian work, the realism and postures of the figures suggest that Audu Mai Alijeta drew inspiration from secondary source material available in England where these gourds were executed. Artists were brought to the London area for the duration of the 1924 exhibition to produce items to sell to visitors. Unfortunately, Audu is not included in a published list of the African artists (Anonymous 1924:55), but the subject matter of his gourds certainly indicates they were made for a non-Nigerian clientele. The techniques used to decorate them, however, are traditional and conform to conventions described above. All the gourds were pressure-engraved with a very sharp point or needle, and the fine execution in Figure 90 is clearly related to one style of pastoral Fulani engraving reproduced here (cf. Fig. 57). Two of the gourds (Figs. 91,92; see Pl. 33) also incorporate chip carving, the Hausa technique used to create a reduced, lightly colored background for the raised motifs (cf. Figs. 61,62a). Both combination of processes and the imagery itself reveal Audu Mai Alijeta's prodigious skill and creativity in gourd decoration.

Despite the above examples, generally there appear to be strong religious reasons why human and animal forms are not represented in the decoration of most gourds. Among the Moslem Hausa, Fulani, and Kanuri there is a clear prohibition

89. Kanuri woman with elaborate coiffure; Maiduguri. 1970.

90 a,b,c. Three views of tubular gourd in Plate 33a, made for the British Empire Exhibition, Wembley, 1924. 36.8 cm. UCLA MCH X65–8232.

against representational art, although in the secular sphere certain allowances have been made. Chappel found that among the animistic Yungur there is a strict taboo against any kind of anthropomorphic imagery on secular objects because such subject matter is strictly reserved for ancestral and healing contexts (1977:126–127). Not only did Yungur elders claim that carvers and their families might suffer harm or even death if they produced such imagery, but any object so decorated would be automatically transformed into a ritual item no longer useful for domestic purposes.

A number of literal iconographic referents have been elicited for gourd designs, and others have been suggested that are highly abstract and capture elusive qualities of things or ideas. For example, idiosyncratic names for designs were collected among the Karekare such as "anger of a stranger," "knees of a talkative woman," or "Muria's eyes."[37] The groups who live in the Uba Plains—the western Margi, the Kilba, and the Chibak—all call the fine line cross-hatching that dominates their design compositions, "beautiful girl." While it would seem that this name is a reference to the geometry and symmetry of the shapes executed in fine lines, one informant—the Chief of Askira, Alhaji Mohammadu Askirama II—claimed the shapes created by the patterning had little to do with a concrete definition of beauty:

> "Did you not know," he asked, "that something beautiful, especially a woman, is something that is smooth and fine? Beautiful women come in all shapes, but what is true about their beauty is also true in the [decorated] gourds—that is, smoothness and fineness to the eye and to the touch."[38]

While the appreciation of gourd decoration cannot be discussed in detail here, it is important to stress that local artists and observers do assess the success or failure of particular compositions.[39] As an example, during the six months Berns spent in the Ga'anda area, she collected a large number of gourds decorated by artists living in different villages. As they accumulated, they were hung on the wall where they could be seen (and admired) by visitors. This method of display did not conform with local Ga'anda custom, but every woman who viewed the collection responded positively to those that were considered *ndican*, a predicate adjective connoting goodness or fineness. Gourds that were *ndican* had symmetrical, balanced compositions whose composite motifs were executed with meticulous precision. Kalhar, a woman recognized as one of the best artists in Ga'anda town, singled out a gourd for approval (Fig. 93a) made by Dije, a highly regarded artist from the town of Boka, whose work clearly met the same high standards as her own (Fig. 93b).[40]

Decorated gourds are preferred over their nondecorated counterparts because they enhance the experience in which they are employed.[41] One young Fulani man told Hudson that being offered water from a girl's gourd with a painted interior constituted an act of flirtation, if not seduction, on her part. Apparently, the more beautiful the gourd revealed itself to be as the contents were drained, the more favorably inclined the young man would be to her, transferring the image of beauty from the gourd to the girl. Accordingly, girls reserve their most beautifully painted gourds for the young men that interest them.

There are some groups, however, who still consider the beauty of a gourd secondary to its utility. For example, among the eastern Margi, the words used to identify a "good calabash" (*ɐncala mɐnagu*) always refer to the gourd's utility, not its beauty (Vaughan 1975:185). Unlike the Ga'anda who have a direct attributive adjective for describing visual characteristics, Vaughan explains that the Margi must go through a circumlocution to express the idea, saying that something is "good to the eye" (*mɐnagu anu li*). Yet, as was proposed earlier in this study, pure considerations of utility make it difficult to explain the time and energy invested in transforming a gourd into a decorative object. Even among the Margi, the fact that women distinguish designs borrowed from other groups reveals a conscious aesthetic awareness. Vaughan notes another important dimension to artistic embellishment that should be considered and may have relevance for other groups as well:

> Decorating calabashes is a social event; women do it as they sit and gossip with their neighbors. Marghi like to do things together; they do not like to be alone, and in some instances it does not seem improbable that being together is more important than the task they seek to complete. I am suggesting here that the social context of decoration is as important as the decorated calabash. I do not mean to say, however, that the decoration is irrelevant, for clearly, Marghi recognize competence and appropriateness in design (1975:185).[42]

It was our observation, however, that many women preferred undertaking this activity on their own. Village life generally precludes extended periods of privacy; women are accustomed to the

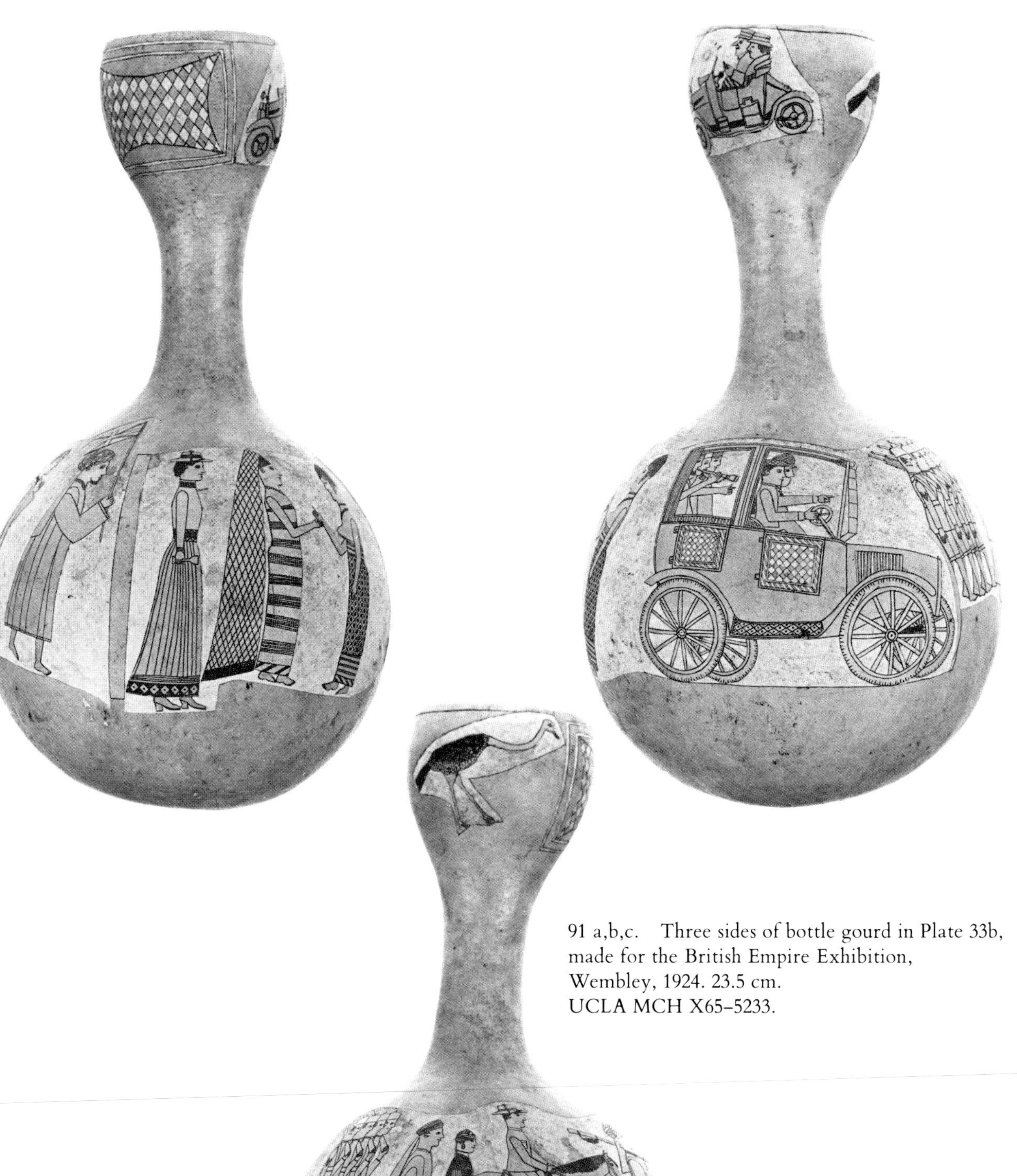

91 a,b,c. Three sides of bottle gourd in Plate 33b, made for the British Empire Exhibition, Wembley, 1924. 23.5 cm.
UCLA MCH X65–5233.

92 a,b,c. Three sides of tubular gourd in Plate 33c, made for the British Empire Exhibition, Wembley, 1924. 38.1 cm.
UCLA MCH X65–5232.

flux of children, relatives, neighbors, and others and do not need a special activity such as gourd decoration to provide a context for socialization. Of the techniques used, however, pressure-engraving is particularly adaptable to social congeniality as it requires no preparation and can be picked up at any time (a factor that makes it so suitable to nomadic peoples). Yet, it should not be forgotten that calabash decoration requires very careful concentration, not only to achieve desired compositions but also to reduce the risk of injuries.

93. Aesthetic comparison of Ga'anda artists.
a. Gourd made by Dije (*tɇb'yata*).
17.8 cm x 21.5 cm. UCLA MCH X85–5;
b. Gourd made by Kalhar (*tɇb'sayema*).
16.5 cm x 18.7 cm. UCLA MCH X85–2.

·4·

THE GA'ANDA

Gourd Decoration from a Sociocultural Perspective

The Ga'anda have been singled out for a detailed case study for two reasons. First, of the 275 gourds comprising the Museum's collection, 75 were made by Ga'anda artists. Second, intensive field research was conducted among the Ga'anda for over six months, yielding information on both the full range of arts produced and the ethnographic context in which they operate.[1] The size of the Ga'anda gourd collection makes it possible to better determine stylistic principles and compositional categories. Accompanying documentation allows the work of particular artists to be identified and, in some cases, to be traced over three generations. Information about context has also made it possible to offer interpretations of designs in more than just a literal referential sense. And lastly, an examination of the other arts produced by the Ga'anda in different media shows that gourd decoration is not an activity with just its own internal rules, but one that belongs to a larger, integrated visual universe.

SETTING

The Ga'anda region is dominated by massed inselbergs and isolated domes that rise dramatically out of an undulating, wooded terrain. Elevations range from 450–600 meters, but summit levels are low compared to the Mandara Mountains to the east. Distinctive geographic landmarks appear to have attracted migrating populations, as settlements were consistently established on the slopes or foothills of prominent hill formations (Pl. 1). The Ga'anda population, dispersed over approximately 250 square kilometers, has moved since the period of colonial reorganization from relatively inaccessible, hilly locations to centralized villages and towns.[2] Ga'anda Town has the largest population and has long been the district's administrative center. Gabun and Boka are the other two sizable towns.

Each Ga'anda locality is an autonomous political unit consisting of related and some unrelated, patrilineal hamlets. Authority is vested in families from which priests and shrine custodians have traditionally been selected. Otherwise, social and economic interactions are governed by the independent decisions of household heads. The main responsibility of local priests (Kutira) is to preside over sacred and ceremonial activities oriented toward spirit worship and social integration. Of all Ga'anda priests, however, the Kuter 'Yera (Rain priest, chosen from the Gudban family) holds the highest position of religious authority. His rank is based on his right to control 'Yera, the spirit force associated with rainfall.

Spirit forces who directly influence health and prosperity are enshrined and worshiped in major religious precincts, *xwer'defta* ("wooded groves"). Ceramic containers objectify the Ga'anda key tutelary forces and are kept in *keten b'uuca* ("houses for pots") erected within sacred groves or individual hamlets. This pattern of ritual decentralization also ensures that each community upholds the social and economic precepts on which the positive intervention of such forces depends.

The Ga'anda topography serves as a natural framework for dispersed settlement. Hamlets are nestled in small clearings in the rocky terrain and include numerous independent households. Marriage is the basis of household organization and is governed by strict rules of exogamy.[3] Marriages are arranged in infancy and are culminated only after a demanding series of reciprocal obligations are fulfilled by the two families. From the earliest stages on, large iron hoe blades (*wanketa*) are given by the family of the prospective groom to that of the bride. The Ga'anda formerly regarded the blades as currency and as primary indicators of wealth and prestige. Large numbers of gourds, pots, and other household items are also given by the groom to his bride-to-be.

The importance of marriage is supported by the rites of transition both Ga'anda boys and girls undergo before they are considered adults. Girls undergo a lengthy and painful program of body and facial scarification, called Hleeta ("scarifying").[4] The sequence is completed in a series of biennial stages, and with each, prescribed areas of a girl's body are cut in increasingly elaborate patterns (Pl. 34). Hleeta also determines the timing for the boy's progressively larger bridewealth payments, which escalate in accordance with Hleeta's successive stages. Boys from six to sixteen are eligible for the initiation ordeal (Sapta) held in Ga'anda localities every seven years.[5] No youth may marry or participate in independent economic pursuits until successfully completing this three-month ordeal.

The Ga'anda are essentially subsistence agriculturalists who farm during a short rainy season (June–October) and supplement their crop yields with game killed during dry season hunts. Farmlands are located some distance from the rugged hills where people live and where little arable land is available. Men and women grow sorghum (*xwerma*), the staple of the Ga'anda diet, and women also plant smaller crops of cow peas, cassava, and ground nuts. In the marginally fertile areas around hamlet centers, women cultivate small plots of tigernuts, bambarra ground nuts, and various leafy shrubs as dietary supplements.

Both men and women engage in one or more part-time occupations during the dry season. All participate in repairing and rebuilding the compound. Blacksmithing, brasscasting, and wood carving are the preserve of skilled male specialists. Any man, regardless of lineage or clan, who proves his aptitude for these skills is eligible to produce the items required by the community. Blacksmithing is a major occupation, and a high demand persists for farming tools and household implements now forged from imported European scrap metal. A number of ornaments and ceremonial hoes, axes, and knives are also produced by blacksmiths for ritual use. Although brasscasting has practically disappeared, Ga'anda was once considered an important center for the production of a wide range of ornaments. Using the lost-wax method, Ga'anda casters made bracelets, beads, bells, knife handles, and small figurines. Brass ornaments are kept by families as traditional attributes of wealth and status and as heirlooms, and they are still worn on ceremonial occasions.

Skilled female artists produce the variety of pottery (*b'uuta*) and decorated calabashes (*njoxtitib'a*) required for domestic and ceremonial use. (A few older men, however, are responsible for modeling the most sacred ceramic vessels.) As mentioned above, large collections of household containers are required as bridewealth. After a bride's marriage, local artists are commissioned to produce the items needed as replacements or as additions to personal collections. Traditionally, qualified women only served the needs of their local community; today, the weekly market system has broadened spheres of social interaction and thereby altered patterns of distribution and consumption. Market traders have also introduced a range of factory-produced enamel containers and aluminum cookware. The trousseau of many modern Ga'anda brides consists almost entirely of these mass-produced items. Despite the availability, popularity, and prestige value of such imported goods, the majority of Ga'anda women still favor the terracotta and calabash containers made according to traditional aesthetic standards. And likewise, their husbands prefer the flavor of food cooked in clay vessels and eating out of gourd bowls.

GOURD PRODUCTION AND DECORATION

The Ga'anda identify at least fourteen different gourd containers, each named according to size, shape, or use.[6] Eight have exteriors ornamented with elaborate pressure-engraved designs and are used primarily for eating and drinking. For bowls, the Ga'anda cultivate a gourd variety that when bisected results in an elliptical rather than a circular section. Decorated spoons are also popular and are made from bottle gourds with long narrow stems (Fig. 94); however, bottle gourds are rarely left whole to be used as containers. Unadorned calabashes are used mostly as carriers or as standardized measures for determining the value of foodstuffs. As described earlier (p. 65), undecorated ovoid-shaped gourds (*sambata*) must be used

94. Array of Ga'anda spoons (*wanb'eleta*). Left to right, from top: a. 19.6 cm x 11.6 cm. UCLA MCH X83–625; b. 15.4 cm x 11.7 cm. UCLA MCH X83–624; c. 21.1 cm x 9.2 cm. UCLA MCH X85–23; d. 23.8 cm x 10.1 cm. UCLA MCH X85–19; e. 21.1 cm x 8.9 cm. UCLA MCH X83–637; f. 27 cm x 11.1 cm. UCLA MCH X85–20; g. 19 cm x 8.4 cm. UCLA MCH X83–634; h. 38.6 cm x 10.7 cm. UCLA MCH X85–26.

for drinking sacred or ceremonial beer (Fig. 30). Ga'anda priests must also eat their food out of *sambata* rather than ornamented containers.

Ga'anda men have exclusive control over the cultivation and preparation of gourds. They are planted early in the rainy season alongside fields of sorghum and are carefully tended until the first fruits appear. Tampering with a man's gourd plants, especially during the early stages of growth, is strongly proscribed, and apotropaic medicines are often placed in the ground around them.[7] Once sufficiently ripe, the new gourds (*tib'a*) are harvested and dried. A special tool kit is used to cut and clean the fruit.[8] To bisect the gourd evenly, a doubled piece of finely plaited fiber cord (*saxtiked'a tib'a*, "cord for marking a gourd") is looped over the cut stem projection and pulled tightly around the gourd. The tip of a curved scraping tool (*wand'ixusta xwertib'a*, "thing for scraping the inside of a gourd") is dragged between the two strands of cord to create a superficial guideline across the circumference (Fig. 95). A clean break is achieved by first sawing a deep incision with a long serated knife (*wand'imbehla*, "a thing for splitting") and then hammering a knife blade (*wand'ikukwata*, "thing for hitting") along the cut with a wood mallet (*d'eftikukwata*, "wood for striking"). Once split, the pulp (*ketere*) is removed and the interior of each bowl is cleaned with the curved blade. Finishing touches include smoothing the rim edge with a potsherd and removing the uneven outer cuticle by scraping or by soaking the gourd shell in water. Ga'anda men once prepared all gourd containers in this way and made them available to a local clientele to be decorated or used as is. Now such gourds are sold at local markets by men who cultivate them for profit.

Any woman may accept commissions for decorating gourds. As with most groups, there is no formal period of training. Young girls learn by watching close female relatives, usually their mothers and grandmothers, and thus inherit the same techniques and styles.[9] Yet, because the Ga'anda are organized patrilineally and live according to rules of patrilocality, qualified artists from the same affinal family may be distributed quite widely. A young girl generally masters the technique before marriage, but a subsequent change of residence (with her gourd collection) would bring her into contact with the work of others. In spite of the potential for diversification, gourds collected among Ga'anda artists and patrons living in widely dispersed communities share features of striking similarity.

95. Farin Jini scraping a guideline around the circumference of an uncut gourd; Ga'anda. February 1981.

96. Ga'anda compositional format, beginning of continuum. Clockwise from top:
a. 16.5 cm x 20.6 cm. UCLA MCH X83–600;
b. 14.6 cm x 21.6 cm. UCLA MCH X83–609;
c. 20 cm x 24.4 cm. UCLA MCH X85–7;
d. 14.9 cm x 18.2 cm. UCLA MCH X83–628.

As described in some detail above, the Ga'anda process of pressure-engraving yields a dense texture of complex linear patterns and a series of bold reserved spaces.[10] Both linear motifs and the negative shapes enclosed by them have names, and there is considerable latitude in how conventional designs are combined on each gourd. Still, the spatial relationship between design elements falls within a continuum defined by two dominant organizational formats: 1) the division of the design field into a number of discrete geometric shapes balanced symmetrically and aligned axially (Fig. 96), and 2) the establishment of an integrated network of shapes that, though symmetrically balanced, are more densely distributed over the gourd surface (Fig. 97). In (and between) both, drawn shapes establish the compositional framework and are characteristically filled with one or more small, densely repeated patterns: rows of zigzags (*wenyiicikan'gera*), zones of checkerboards (*wanfiyca*), and chains of diamonds (*lan'taya*).[11] A final filler motif called *saxta*—the name given to the sets of parallel lines filled with cross-hatchings (*cerahla*)—is the most commonly used. Strict alternations of *saxta* "ropes" encircle the gourd about 2 cm from the edge, serving as the primary framing device. Bands of *saxta* also delimit geometric shapes that either generate further geometric shapes or are filled with one or more of the other textured patterns listed above (Fig. 98). When *saxta* ropes are aligned as a series of juxtaposed chevrons, they are called *ʔaliyirfa* ("fishbones"; cf. Fig. 98b,f).

At one end of the continuum (i.e., the first compositional format), the spatial relationship between discrete elements is controlled and formal. The inscribed figures float against a neutral ground, and their placement sometimes establishes a striking secondary pattern in the unmarked shell (Fig. 96). Dyeing a gourd red also accentuates the contrast between smooth, untextured shapes (*lifo*, "strips of goat skin") and blackened registers of incised design. Figure 96 shows two devices Ga'anda artists use to maintain a formal compositional balance: aligning identical motifs along the same axis and engraving adjacent motifs with mirrored contours. The strict regularity and symmetry of filler pat-

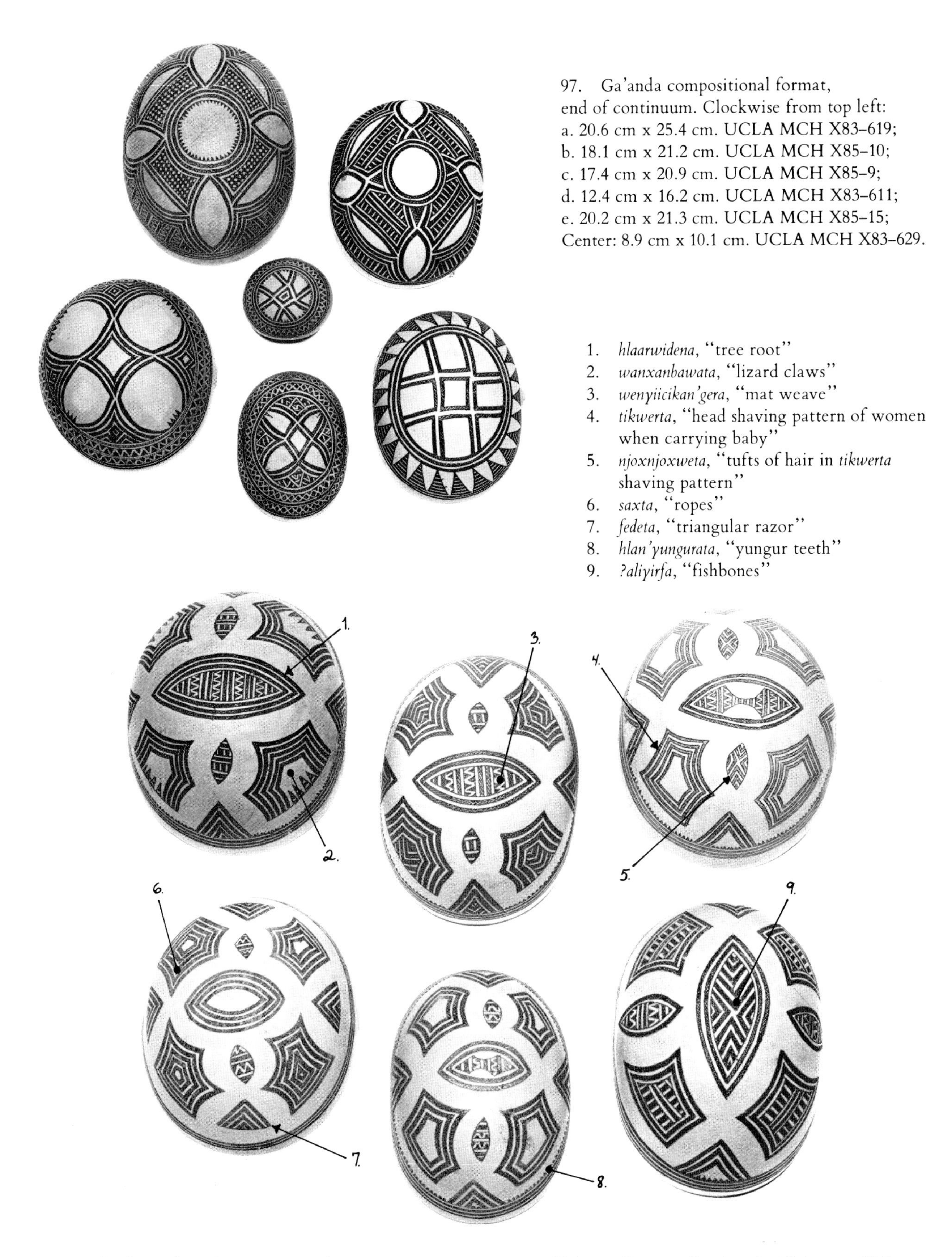

97. Ga'anda compositional format, end of continuum. Clockwise from top left: a. 20.6 cm x 25.4 cm. UCLA MCH X83–619; b. 18.1 cm x 21.2 cm. UCLA MCH X85–10; c. 17.4 cm x 20.9 cm. UCLA MCH X85–9; d. 12.4 cm x 16.2 cm. UCLA MCH X83–611; e. 20.2 cm x 21.3 cm. UCLA MCH X85–15; Center: 8.9 cm x 10.1 cm. UCLA MCH X83–629.

1. *hlaarwidena*, "tree root"
2. *wanxanbawata*, "lizard claws"
3. *wenyiicikan'gera*, "mat weave"
4. *tikwerta*, "head shaving pattern of women when carrying baby"
5. *njoxnjoxweta*, "tufts of hair in *tikwerta* shaving pattern"
6. *saxta*, "ropes"
7. *fedeta*, "triangular razor"
8. *hlan'yungurata*, "yungur teeth"
9. *?aliyirfa*, "fishbones"

98. Ga'anda preferred compositional arrangement: a. 19.5 cm x 23.2 cm. Berns collection; b. 17.4 cm x 22.8 cm. UCLA MCH X85–6; c. 20 cm x 23.2 cm. UCLA MCH X83–610; d. 18.6 cm x 20.9 cm. UCLA MCH X83–622; e. 15.2 cm x 20.3 cm. UCLA MCH X83–620; f. 17.5 cm x 23.8 cm. UCLA MCH X83–616. All of the above, with the exception of "d," were decorated by the artist, Kalhar.

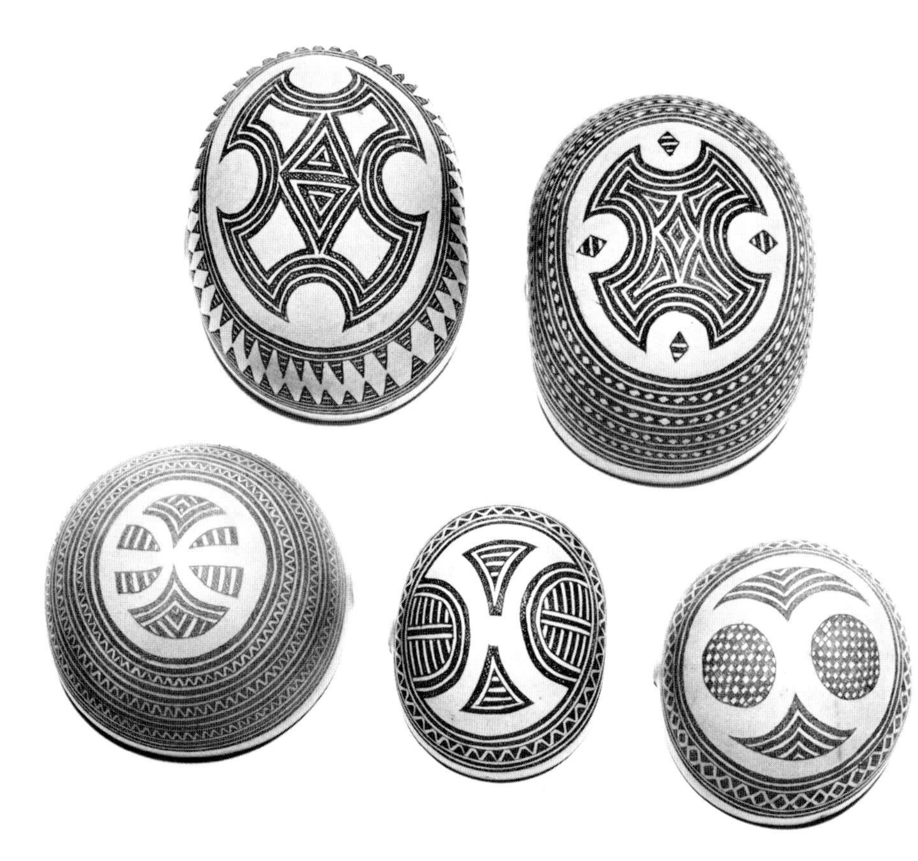

99.
Ga'anda compositional format, middle of continuum:
a. 17.8 cm x 22.8 cm. UCLA MCH X83–606;
b. 17.5 cm x 23.5 cm. UCLA MCH X83–632;
c. 18.1 cm x 17.9 cm. UCLA MCH X85–17;
d. 12.7 cm x 15.5 cm. UCLA MCH X83–623;
e. 15.2 cm x 16.2 cm. UCLA MCH X85–11.

terns intensify the order and unity of each composition. Names identifying elements of a composition reflect an awareness of these basic stylistic features—they differentiate the shapes drawn, the configuration of the spaces left between them, and the patterns used to fill them. In Figure 96c, the entire design is named after the grid formed by the radial placement of alternating triangles and rectangles, *njarta mbanda* ("crossroads"). The triangles also are named, *fedeta* ("triangular razor"), and each shape is defined by and filled with *saxta* ropes.

Most compositions classified in this first category are based on a simple disposition of geometric shapes. There is, however, one rather complex arrangement preferred by a few artists (Fig. 98). Five of the six examples illustrated here were engraved by the well-known Ga'anda artist, Kalhar.[12] She freely interpreted the same conventional format by altering the size of the composite motifs and the filler patterns enclosed by them. Figure 98f demonstrates how a vertical, rather than a horizontal, alignment of the central oval accentuates the ovoid shape of the bowl; it also creates a distinctive pattern in the reserved shell.

Further along the continuum, Ga'anda gourd designs become more complex and have a greater number of compositional variations. This is reflected in the specificity of named motifs, such as those in Figure 99e. There, edges are framed with alternating bands of *saxta* and *lan'taya* diamonds, and the central design field is dominated by two repeated shapes, each filled with a different pattern: circular "hair tufts" (*pwoti*) are filled with checkerboards (*wanfiyca*), and triangular "markings on a donkey's back" (*pelfede kwaarii*) are filled with *saxta* worked in a "fishbone" (*?aliyirfa*) pattern. Generally, there is a more pronounced distinction between the center and the marginal areas of the design field. As Figure 99 shows, motifs are organized around one bold central design, such as an arrangement of one or more circular hair tufts (*pwoti*; Fig. 99e), the distinctive curvilinear shape of a wooden knife hilt (*?inhluuta*; Fig. 99d), or the scallop-shaped scarification pattern under a girl's navel (*kun'kanwannjinda*; Pl. 35 and Fig. 99a,b). These are surrounded by registers of secondary designs that continue around and down the sides of the gourd: small triangular charms (*tiltil*; Fig. 99b), rows of *saxta* ropes (Fig. 99a-e), and small *wenyiici-kan'gera* zigzags (Fig. 99d).

100. Ga'anda beer bowl (*tɇb'kɇnnda*). 18.4 cm x 27.5 cm. Berns collection.

Distinctive to the most integrated compositional formats is the way reserved shapes tend to be framed by dense filler patterns (Fig. 97). This orientation results in even more distinct figure/ground inversions than those evident in the preceding categories of Ga'anda gourds. This is probably nowhere more evident than in the distinctive compositional variations illustrated in Figures 97 and 100. In the first, the central inscribed circle, *pwoti* ("hair tuft"), is framed by the radial placement of oval "leopard eyes" (*wenitipekta*) that emerge as figures against a ground of *saxta* ropes and other filler motifs (Fig. 97a,b). In Figure 100, the central *pwoti* is surrounded by wide radial bands filled with alternating chains of *lan'taya* diamonds and *saxta* ropes. Between these four main units are distinctive sawtooth shapes that emerge in reverse relief because of the illusion created by the background of *saxta* ropes. That this spatial inversion is recognized is confirmed by the fact that in the latter example the unworked shapes are named, *wanxanbawata* ("lizard claws"), and the triangles between the lizard's nails are named separately after similarly shaped brass charms (*tiltil*). A related gourd, decorated about ninety years ago by the grandmother of a highly reputed Ga'anda artist named Witebar, varies somewhat in choice of filler motifs (see Fig. 15a).[13] Nevertheless, it shows that the same conventional combination of designs has been used by at least three generations of Ga'anda engravers.

In the above subcategories the interpretive variations in gourd design are impressive, ranging from subtle to dramatic. It is clear that while artists seem to prefer certain conventional combinations, the range of filler motifs allows even rigid decorative formats to be manipulated creatively. The same is true even on the limited design field on Ga'anda gourd spoons (Fig. 94).

ARTISTS AT WORK

The technical restrictions implicit in pressure-engraving make the range of compositional diversity and creativity evident in Ga'anda gourds truly outstanding. Because the Ga'anda prefer ovoid bowls, this means the gourd is usually oriented to coordinate the composition with a longer vertical axis and a shorter stem-base axis (see Fig. 98). These formal imperatives help explain the expediency of quadrapartite compositions comprised of geometric units arranged radially and multiplied in a more or less complex fashion.

Detailed examinations of how Ga'anda artists work provide some insights into the harmonious marriage of design and technique in creating gourd compositions.[14] Figure 101 shows how Adio Ga'anda built the "discrete" shapes of the first organizational format.[15] She divided the field symmetrically by placing four diamonds at either end of the horizontal and vertical axes (determined by the position of the stem blemish). The area between each pair of diamonds was then filled with identical secondary motifs, leaving a space for a final central diamond. Once this composition was blocked out, the shapes were filled with patterned motifs, and as the illustration shows, Adio completed each visible zone before rotating the gourd further.

A second gourd engraved by the same artist illustrates how one of the more complex, integrated design compositions was constructed (Fig. 102). The central design field was established by drawing an outer boundary of framing bands parallel to the rim and an inner circle around the top of the gourd. As the diagram shows, Adio then divided the field into four sections by axially placing repeated ovals (*wenitipekta*, "leopard eyes"). The complex motifs that fill each identical division are also completed sequentially.

A third gourd, decorated by Fadimatu Ga'anda, shows a design composition built to achieve a figure/ground reversal (Fig. 103). Unlike the first two examples, this gourd is small (15 cm), which allowed Fadimatu to see more of its surface, as well as to rotate it more easily. In fact, the kind of composition used here, based on successive registers of encircling lines, is generally confined to gourds small enough to handle in this way. As the diagram shows, even the series of small ovals (*wenitibed'eta*, "eyes of locusts") drawn between the lines were executed as a continuous series of short curves impressed evenly around the gourd in one direction and then completed with a second set curving in the opposite direction. The short horizontal lines engraved between the ovals were filled with hatchings to create the inverted ground. Diagramming an artist's work shows that while the central circle of the gourd appears dominated by a drawn cloverleaf pattern, this is really the space remaining after the even placement of four curvilinear triangles, *pelfede kwaari* ("donkey back markings"), filled with *saxta* ropes.

The last gourd diagrammed here was decorated by Dije, an artist living in the town of Boka (Fig. 104; see also Fig. 54). Like the gourd made by Fadimatu of Ga'anda, this relatively small example demonstrates how an artist can effectively create continuous lines with a series of economical movements. As shown, sets of parallel lines were completed by working in progressive vertical segments, each equivalent to the length of one engraved stroke. This approach frees the artist from rotating the gourd with every movement and helps maintain the even spacing between lines. Once this circular framework is blocked out, small "filler" motifs are meticulously drawn. Figure 104 shows how tight chains of chevrons (*wenyiicikan'gera*) were impressed within each of the vertical segments. *Cerahla* hatchings are cut similarly and the rigid position in which the gourd is held defines the scope within which an artist can most efficiently work. Dije was commissioned to execute the unblackened gourd illustrated in Figure 54 to show how *njoxta* designs emerge in relief after *cerahla* hatchings have been completed. Although blackening would intensify the contrast between worked and unworked areas, a detailed view of the surface shows how much of this effect is achieved through the laborious process of pressure-engraving alone.

The second stage of *cerahla* is the most time-consuming, and artists rarely complete it in one sitting. The *njoxta* outlines that precede it are done more quickly. For example, Adio completed two *njoxta* gourd compositions in just under an hour; Fadimatu engraved designs over a smaller field in just over thirty minutes. Dije completed the *njoxta* and filled in some of the *cerahla* cross-hatchings on the gourd illustrated in Figure 54 in under one hour. Dije explained that customarily she only executes the first stage of engraving (*njoxta*) for her patrons, who then complete the tedious cross-hatchings themselves (as well as apply charcoal and oil to reverse the designs). Not all artists work in this way nor are all Ga'anda women able to do *cerahla*, but this approach allows an artist to accept more commissions. Dije can finish the *njoxta* compositions on two larger gourds or five smaller ones in a day, in addition to meeting her household obligations. It may be for the same reason that Dije has begun filling larger shapes with cross-hatched lines, emulating the quicker, broader strokes of pyro-engraving (Fig. 67c).[16]

While it may be difficult to recognize the hand of a particular artist, Ga'anda women can identify the work of those who are especially well-known. Dije, for example, is known for her work both by women living in Ga'anda, where her mother now

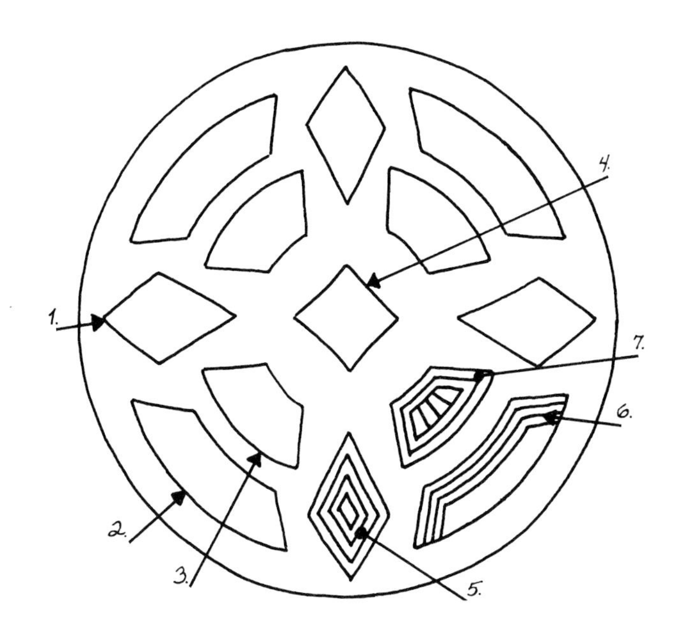

101. Diagram of how a design composition by Adio Ga'anda is constructed.

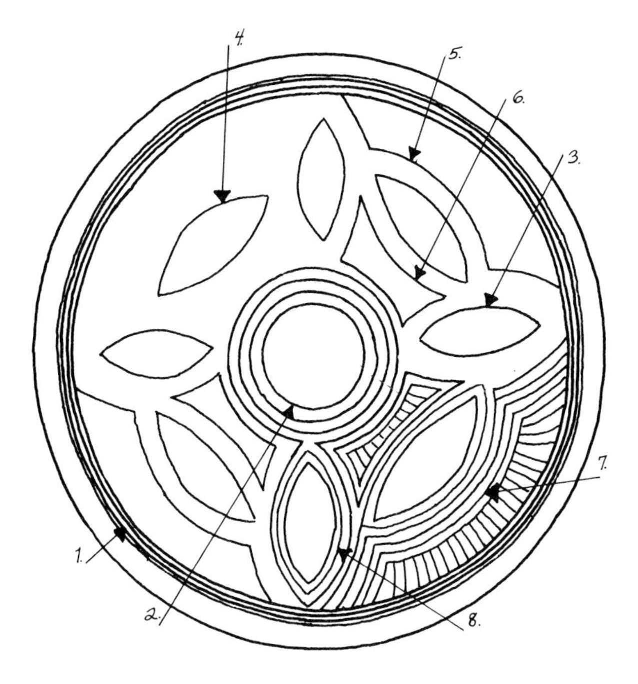

102. Diagram of how a design composition by Adio Ga'anda is constructed.

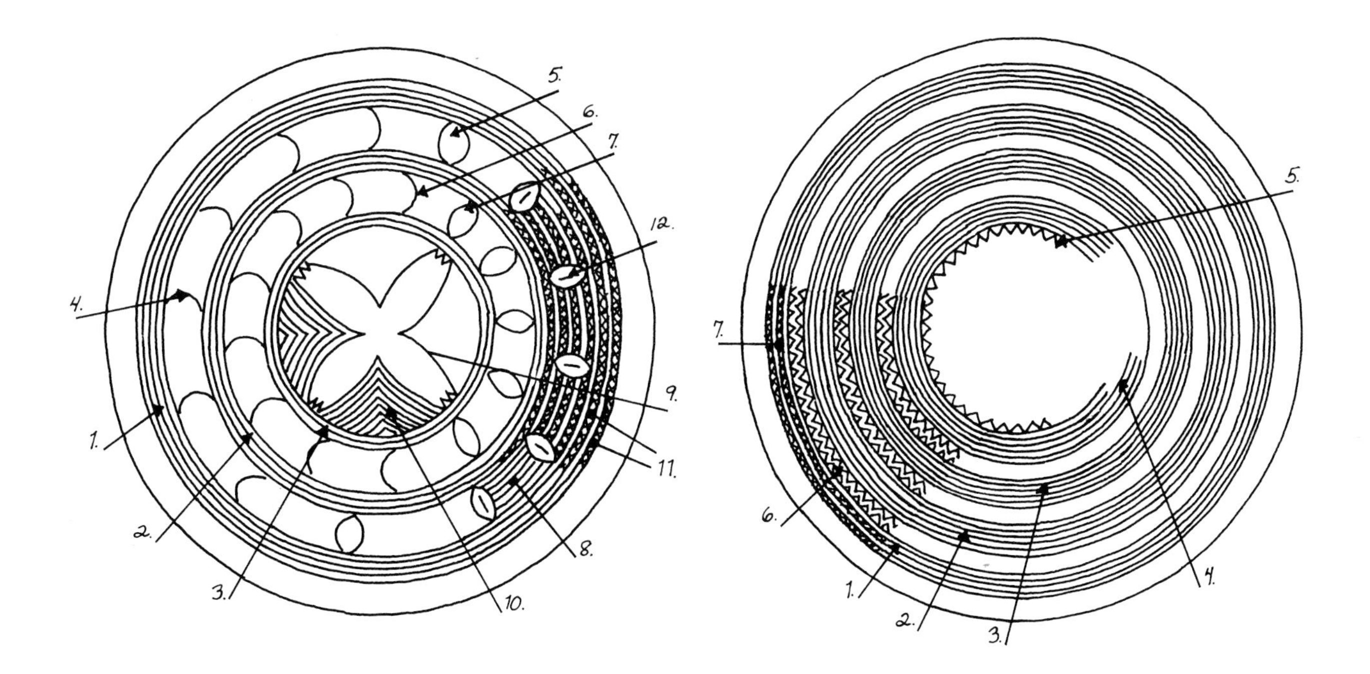

103. Diagram of how a design composition by Fadimatu Ga'anda is constructed.

104. Diagram of how a design composition by Dije Boka is constructed.

105. Dije applying blackening to an engraved gourd spoon (*wanb'eleta*); Boka. February 1981.

106. Kalhar pressure-engraving a gourd bowl (*tib'a*); Ga'anda. 1970.

lives and in Boka, where she was born and still lives with her husband (Fig. 105). Her gourds were also seen by Berns in a village seven kilometers away (Sama) where a Boka girl had moved after marriage. Gourds by Dije in the Museum's collection include examples of both her earliest and most recent work (Fig. 15b,c).[17] A comparison of the two large ovoid gourds, of the sort Ga'anda women use to serve their husband's morning beer, shows what Dije claims to be a technical and compositional improvement over an early stage in her career.

While Dije is an accomplished artist, Kalhar (mentioned earlier) is recognized as being one of the best in her community (Fig. 106).[18] Examples of her work were collected by Hudson in 1970 and by Berns in 1981, and each reveals Kalhar's consummate talent for pressure-engraving (Fig. 98). Even in the smallest spoons, she consistently maintains an extreme regularity in banded alternations and filler patterns, a delicate balance in figure/ground spatial relationships, and an overall symmetry. The same excellence in execution and composition are evident in Figure 97a, a gourd decorated by Kalhar's mother around 1940.

THE INTEGRATION OF GA'ANDA ARTS AND THEIR MEANINGS

While a study of Ga'anda gourd decoration could progress in a microcosmic direction—focusing specifically on artists' personal histories or on a precise comparison of work done from community to community or even hamlet to hamlet—there is more to be gained by exploring how this activity fits within the larger Ga'anda artistic macrocosm. As suggested briefly in the ethnographic background provided above, the Ga'anda produce a range of objects in different media that is integrated into their social and spiritual life. The similarities between categories of art made by Ga'anda-speaking peoples are especially remarkable in light of the differences one might expect to find based on their distribution in pockets of rigorous terrain and their diversification into three dialect sections (Ga'anda, Boka, and Gabun). The considerable creative license found in gourd decoration is also evident in most other categories of art. Yet, as demonstrated below, consistent structural principles govern the evolution of a highly integrated visual system. Their delineation helps explain the expressive role of a wide range of Ga'anda arts, including decorated gourds.

What is classified here as "art" represents a complex corpus of object types originally conceived to facilitate basic economic processes. Granary construction, pottery making, gourd preparation, and blacksmithing are all fundamental to an agricultural society. The essential domestic value of containers and tools produced by these processes provides a basis for their translation into objects of social and sacred importance. Despite artistic transformations, these forms are still fundamentally oriented toward meeting the demands of daily life. However, their ideological value has been enhanced by the imposition of designs that identify and define their contextual role. Public arts, which encompass gourds and domestic pottery as well as ornaments worn on festival occasions, are decorated with motifs that communicate the importance of basic social and economic precepts. Likewise, ceramic vessels made by the Ga'anda to foster and focus spiritual interaction rely on patterns of surface alteration to differentiate the identities and roles of the forces they contain.

Both secular and sacred arts reveal decorative convergences that strengthen their individual value as visual communicators. Consistencies are evident in their decorative vocabularies, in the formal juxtapositions of motifs, and in the processes used to execute them. As with gourd decoration, an ordered repetition of a limited number of simple geometric shapes tends to dominate all design programs. Most distinctive about this particular repertoire of motifs is how closely it correlates with the designs that constitute the Ga'anda program of body and facial scarifications (Hleeta; Fig. 107). Their configurations and their names are often the same and suggest the possibility that Hleeta may have been the visual source for many designs transposed into other media. For example, *caxi'yata* curves at the abdomen, *kun'kanwannjinda* scallops under the navel, and *kwardata* lozenges down the thighs and the sides of the torso are common motifs. They are evident not only in the iconography of gourd decoration (Figs. 108,109), but also in the bold designs painted across mud granaries and the delicate patterns on cast bracelets (Pl. 36; Fig. 110b,c,d). Because Hleeta is completed in six stages, the controlled alteration of a girl's body allows discrete units of decoration to remain isolated for extended periods of time (cf. Pl. 34).[19] Indeed, it may be that the message implicit in Hleeta is contained in each of its distinctive composite parts. It is proposed here that the meticulous imposition of designs on a woman's body may have established a pattern and provided a model

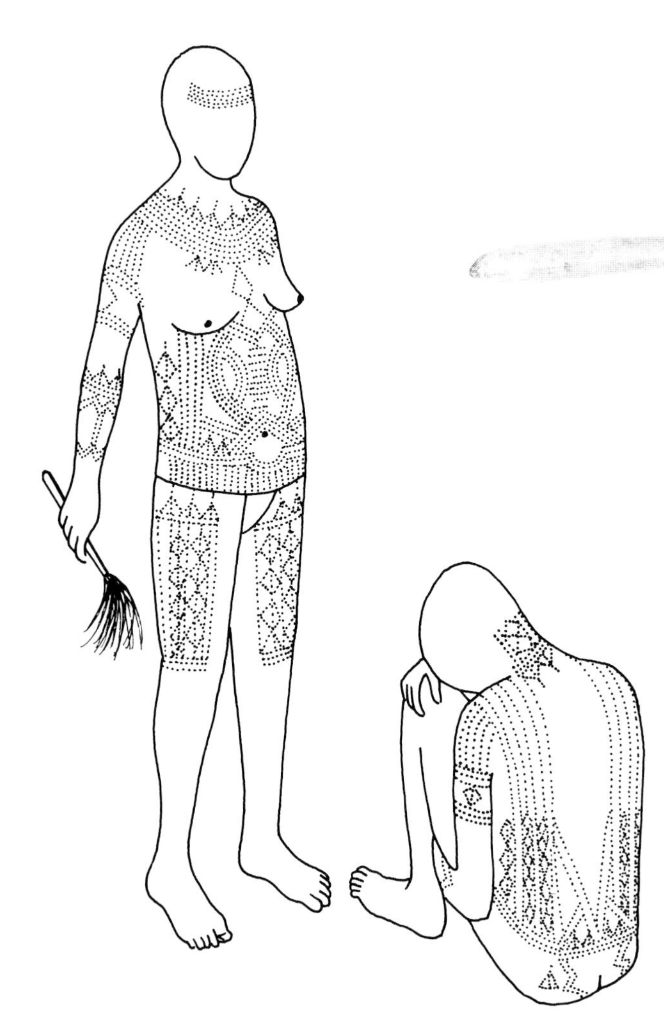

107. Ga'anda scarifications (Hleeta);
figure contours drawn after Chappel (1977:fig. 222).

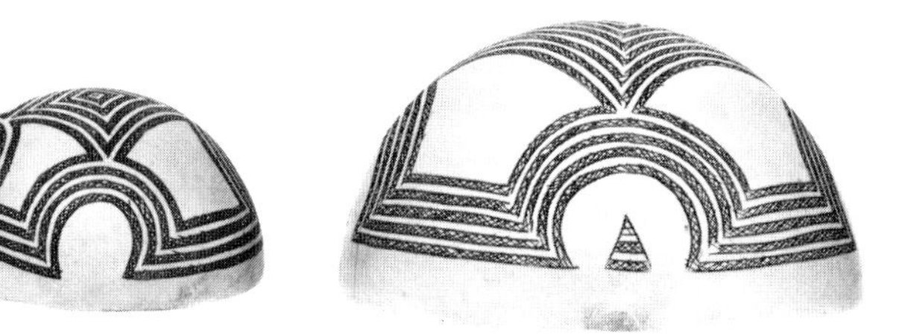

108. Ga'anda decorated gourds with scallop motif named after scarification markings worked under a woman's navel (*kun'kanwannjinda*):
a. Spoon. 24.7 cm x 12.7 cm. UCLA MCH X85–21;
b. Bowl. 16.5 cm x 19 cm. UCLA MCH X83–605.

109. Detail of Figure 100 showing the lozenge-shaped motifs worked along a woman's torso and thighs (*kwardata*).

110. Ga'anda cast brasses.
a. Knife hilt (*?inhluuta*).
9.5 cm x 6.7 cm.
UCLA MCH X85–107;
b. Armlet (*ro?hlon'nda*).
11.7 cm x 8.3 cm.
UCLA MCH X85–108;
c. Armlet (*ro?hlon'nda*).
12 cm x 8.3 cm.
UCLA MCH X85–98;
d. Child's armlet.
8.6 cm x 7 cm.
UCLA MCH X85–99.

for altering the surfaces of other objects used in related contexts. Moreover, the angular shapes that dominate in Hleeta appear to have become primary symbolic emblems of marriage and social stability. The transposition of Hleeta designs into other media also suggests that aesthetic choices are governed, at least in part, by their social utility and communicative value.

Calabashes are an important part of each girl's bridewealth, and their decoration enhances their role as expressive emblems of domestic rights, obligations, and capabilities. While some gourd motifs resemble and have the same names as scarification patterns, the decoration of other Ga'anda arts have designs that refer even more directly to the message of Hleeta. The most visible example of how a decorative program is manipulated to exploit its communicative potential is the special compound, *kɇtɇn perra* ("new bride's house"), that a husband must build before his marriage can be finalized. The compound is constructed according to conventional Ga'anda architectural specifications, but a series of decorative attachments differentiate the woven grass facade from other hamlet structures. The most notable of these is a continuous row of inverted triangles or lozenges tied along the outer wall with ropes; these forms are plastered with a mixture of mud and red ochre (Fig. 111). The shape and color of this bold mud frieze seem to refer directly to the bodies of young women at the festivities following the completion of Hleeta; they are marked with predominantly angular rows of cicatrices and are smeared with oil and red hematite. These women also wear cast brass triangular charms (*tiltil*) lashed onto bands of reddened goatskin tied across the chest (Fig. 112). Thus, the series of concise symbolic markers on the compound publicly announces the establishment of a new patrilineal household through an obvious reference to the bride's legitimate status.[20] The association of this graphic tonal and textural syntax with brides is repeated in the wickerwork baskets (*can'lan'nda*) used for displaying a girl's gourd bridewealth (Pl. 7). Over a bent wood armature, braided rope of natural and dyed fibers is wrapped to emulate the distinctive zigzag lintel tied across the facade of the bride's house. Over this, strips of smooth red-dyed goatskin are crisscrossed to create a continuous pattern of linked lozenges.

Ga'anda men further acknowledge the economic importance of their wives through the bold designs painted on the large granaries that occupy the public sectors of each hamlet (Pl. 36). As containers of seasonal agricultural harvests and emblems of household prosperity, granaries are appropriate and conspicuous armatures for the imposition of socially meaningful (as well as aesthetically pleasing) designs. The example reproduced here was painted with an alternating pattern of diamonds and circles superimposed over a series of vertical stripes that divide the design field into zones. Both the shapes and their juxtaposition correlate with motifs from Hleeta (cf. Fig. 107). Also notable is how the patterned alternation of interlocking shapes maintains a design principle consistent with calabash decoration.

Because designs used on these structures refer specifically to brides, they reinforce the particularly important place of women in Ga'anda household economics.[21] The time and effort invested in the decoration of gourd bowls is one way of acknowledging the pivotal role of women in food preparation. The way calabashes project meaning, however, may be enhanced by how their decoration correlates with that of other arts oriented toward the same ends (e.g., Fig. 113). Ritual pots, such as *wantafata* and *b'uutiwankɇta*, used by the groom to make substantial payments in beer to the bride's family, have motifs associated with the transition of youth to adulthood (Figs. 114,115). The shapes, as well as the names of the designs boldly drawn over the rouletted shoulder of the *b'uutiwankɇta* (Fig. 115) refer directly to scarification—*kwardata* (lozenges) and *caxi'yata* (curves). They are also found on ornamental iron throwing knives (*cicawa*) carried prominently during the final celebration of a boy's graduation from the septennial initiation ordeal (Sapta; Fig. 116). Not only are these knives hammered with rows of linked lozenges, but some examples also have checkerboard grids (*wanfiyca*) and juxtaposed chevrons (*?aliyirfa*) that recall patterns distinctive to gourd decoration (see Fig. 122a). Additionally, the projecting wedge-shaped flanges of cast brass armlets (*ro?hlon'nda*), worn both by boys and girls on festival occasions, are decorated with related patterns of lozenges and vertical lines (Fig. 110b-d).[22]

The repetition of a limited universe of motifs is not the only device that projects meaning. Alterations of surfaces may also create distinctive textural contrasts by which social and spiritual processes are acknowledged and celebrated. Hleeta scarifications and their extensions, rearrangements, and combinations demonstrate how surface alterations can transform objects and infuse them with meaning. The tiny cuts made in the skin create an even pattern of raised scars whose visual impact is determined by subtle relief as well as by tactile

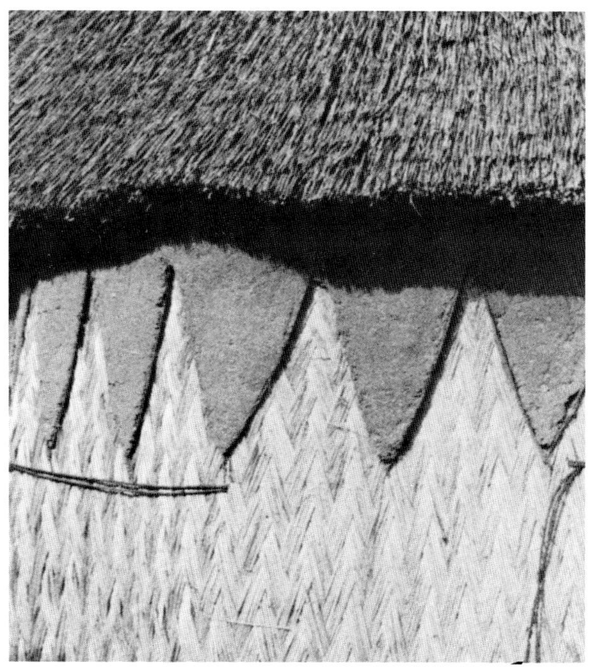

111. Detail of the mud frieze worked on the facade of a bride's compound (*keten perra*); Kwanda. January 1981.

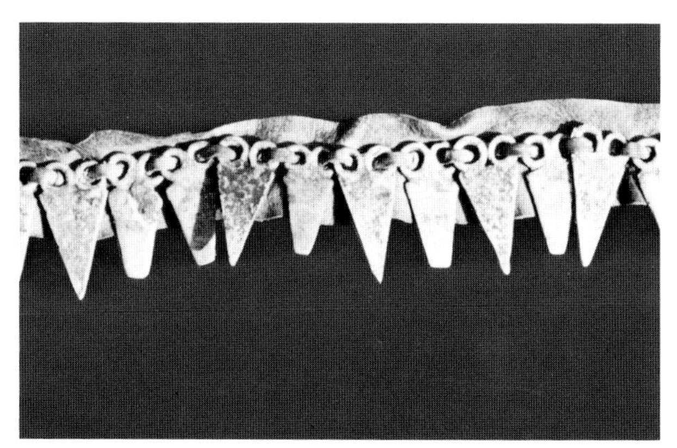

112. Ga'anda cast brass triangular pendants (*tiltil*) threaded on a leather strap. 4 cm. 1981.

113. Detail of Figure 71. Ga'anda gourd bottle (*buta*) with *tiltil* motifs.

114. Detail of Ga'anda ritual beer vessel (*wantafata*) with *tiltil* motifs. Ceramic. 50 cm. February 1981.

115. Detail of a Ga'anda ritual beer vessel (*b'uutiwanketa*) with *kwardata* and *caxi'yata* motifs.
Ceramic. 85 cm. February 1981.

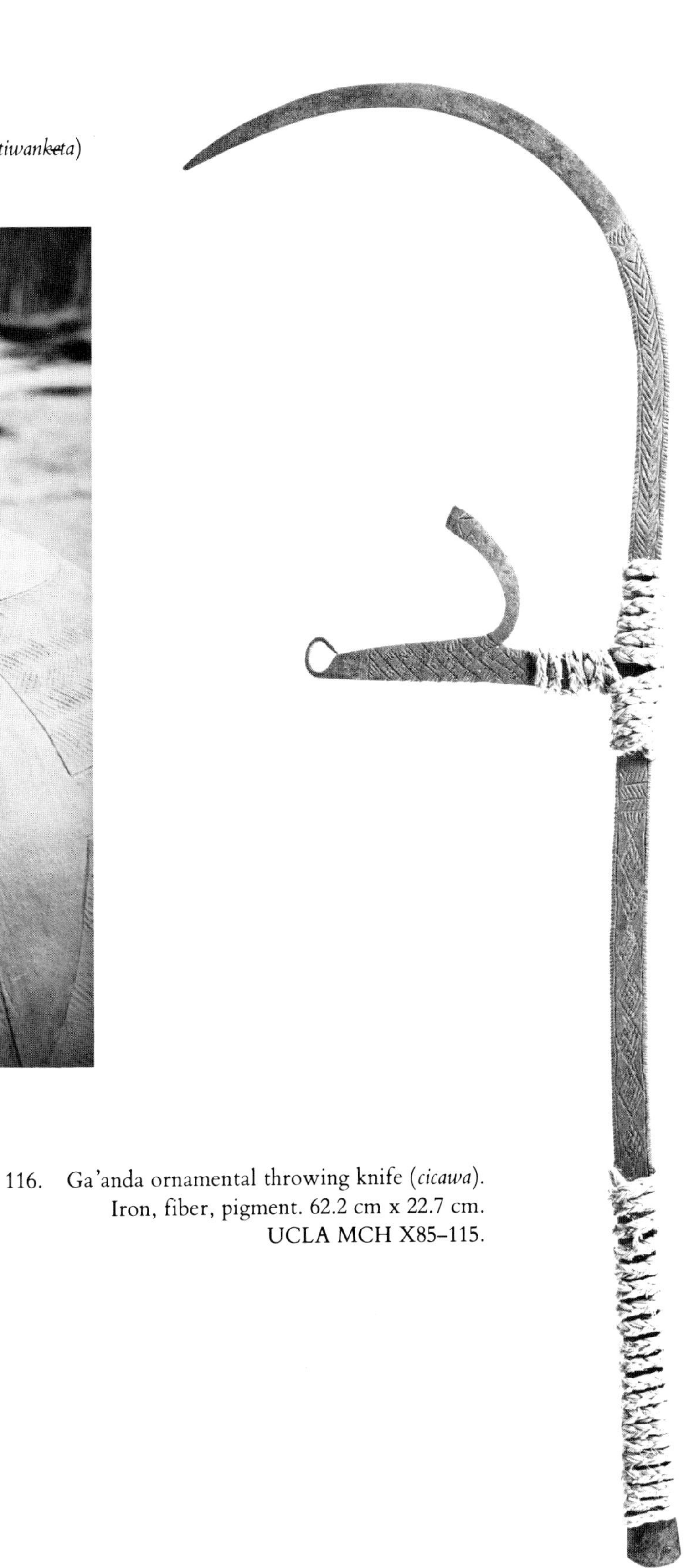

116. Ga'anda ornamental throwing knife (*cicawa*).
Iron, fiber, pigment. 62.2 cm x 22.7 cm.
UCLA MCH X85–115.

contrasts (Pl. 34; Fig. 106). The primary designs also delineate areas of unmarked skin that emerge as secondary shapes. For example, when *kwardata* lozenges, *caxi'yata* curves, and *kun'kanwannjinda* scallops are transposed into other media they quote both the linear alignment of cuts and the planar shapes they outline.

The same principle operates in calabash decoration where compositional formats stress a complex interplay between engraved linear motifs and the smooth reserved spaces between them. As described above, the banded shapes that float against a neutral, unmarked ground are manipulated to create emphatic secondary patterns in the reserved areas. At the same time, the engraving process can reverse this figure/ground relationship: reserved shapes can be drawn so they appear superimposed against a background of densely incised motifs (Fig. 113).

The same spatial flexibility is evident in pottery decoration. Rows of dense, regular calabash or stalk impressions outline shapes that transform the planar surfaces around them into a second design pattern often accentuated by burnishing (Fig. 115). Furthermore, bold geometric shapes can be incised against a rouletted ground to create the same spatial inversion seen on calabashes. This latter illusion is demonstrated clearly in the way rows of smooth tapering triangles (*tiltil*) are imposed against a textured ground in both media (cf. Figs. 113,114).

The significance of particular shapes and combinations of textures is apparent in the ways Ga'anda spirit beings are objectified in ceramic form. Ngum-Ngumi, the spirit responsible for leading the Ga'anda in their migration from the "East" and for continuing to protect them, is best identified by a central panel of dense corrugations, each meticulously impressed with the straight edge of a calabash chip (Fig. 117, back). A second powerful protective presence is Mbirhlen'nda, whose decorative features are dominated by a diagnostic network of nodes applied to the head and neck, ending in a collar of large triangles (Fig. 117, front).[23]

117. Ga'anda sacred vessels enshrined in Pitel with Ngum-Ngumi in the back (also called Yinifarihi) and Mbirhlen'nda in the front. Ceramic. 65 cm and 52 cm. January 1981.

This same syntax of juxtaposed textures dominates most categories of Ga'anda art, strengthening the visual coherence of particular geometric shapes. For example, *kwardata* lozenges hammered along the planar blades of dance axes (*wurta*) are smooth shapes outlined and emphasized against a ground of densely hammered lines. Likewise, the motifs etched into the wax molds used in casting knife handles (*?inhluuta*) are consistently oriented against a network of fine scoring (Fig. 110a). The same contrasts can be identified in the structure and decoration of large mud granaries belonging to women. The platform is often built of closely aligned, small mud units painted in alternating patterns to emphasize regularity and symmetry (Pl. 36). Additionally, the visible upper flange may be decorated with a continuous frieze of lozenges or triangles drawn against a textured ground, identical in syntax to the way other surfaces are treated.

Such consistencies in shape delineation and textural contrast that characterize the decorative programs of both secular and sacred Ga'anda arts cannot be separated from the processes used to achieve them. The convergences in technical constraints are equally intriguing, despite the range of media involved. Hleeta scarifications are accomplished by carefully piercing and lifting a small ridge of skin with an iron hook and then quickly cutting across it with a sharp razor. This painstaking (and painful) procedure requires the artist to concentrate on limited areas of the body at a time. The result is a controlled and progressive accretion of restrained design units. As described in some detail above, the laborious process of pressure-engraving imposes the same constraints, as the artist uses short, controlled strokes to cover small areas. Like scarification, compositions are not broadly conceived, but represent an accumulation of balanced and integrated units. The Ga'anda preference for dyeing gourds red accentuates the contrast between smooth and engraved areas. This correlates in an interesting way with the results of applying oil and red hematite to a young woman's body—it too greatly enhances and makes visible the subtle texture of raised scars against the skin.[24]

The same restraint and precision are exercised in creating textures in clay and mud sculpture, although the plastic advantages of both media would allow much greater decorative flexibility. Rows of densely placed calabash chip or split stalk impressions cover the neck sections of all Ga'anda domestic wares. They are painstakingly worked around a vessel in progressive vertical columns (Fig. 118), an approach that echoes the way Ga'anda women engrave "continuous" lines around the surface of a gourd (cf. Fig. 104). This working procedure can also be compared to the way ridges of skin are lifted one at a time before being carefully scarified in tight rows.

118. Dije decorating the neck of a Ga'anda water pot (*b'uuti'yema*) with a calabash chip (*caxa*); Boka. Note the scallop motif on the pot body (*kun'kanwannjinda*) named after a woman's scarification motif. February 1981.

The surface modeling of the two primary sacred vessels—Ngum-Ngumi and Mbirhlen'nda—also relies on laboriously executed relief textures (Fig. 117). The way most women's mud granaries are constructed even more dramatically reveals the accommodation of aesthetic preferences to a set of technical determinants (Pl. 36). On the one hand, the complex openwork platforms provide an even support for the combined weight of the broad mud chamber and its contents; they also increase ventilation and help protect foodstuffs from vermin and domestic animals. On the other hand, their sculptural elaboration reflects a desire to achieve a composition of small, regular, repeated units.

The controlled juxtaposition of geometric motifs and the careful execution of complex textures dovetail in an intriguing way with designs that result from certain basic technological processes employed to produce the structures that dominate the Ga'anda landscape and to make other domestic items. The named vocabulary of filler designs recorded for calabash decoration identify clearly the graphic transposition of techniques, like weaving, knotting, and braiding: *wenyiicikan'gera*, a triangular interlace pattern that duplicates "palm frond mat weave" (Fig. 119); *lan'taya*, a continuous string of openwork lozenges that represents "knotted fish net" (Fig. 120); *saxta*, the bands of cross-hatched texture that re-create lengths of "braided rope" (Fig. 121); and *wanfiyca*, a checkerboard grid that is a stylization of the dark/light variegation and the interlocking weave of "royal cloth aprons" (Fig. 122). The methodical process of pressure-engraving lends itself well to the two-dimensional reinterpretation of designs that are by nature tight, regular, repetitive, and fundamentally linear.

Complementing such highly textured patterns are organic shapes that demonstrate an awareness of designs inherent in nature: the markings on a donkey's back (Fig. 99e), the grid motif on a turtle's shell (Fig. 97c), the oval eyes of a leopard (Fig. 97a,b), the claws of a lizard (Fig. 100), or the circular eyes of an insect (Fig. 103). These calabash motifs share the same conceptual feature—they isolate a visually distinctive element that differentiates one organism from another. Even the patterns that distinguish the structure of natural forms are identified in the Ga'anda design vocabulary—fish bones, tree roots, and forked branches. In addition to the "process" designs mentioned above, the Ga'anda also associate a number of geometric shapes with other distinctive forms of cultural expression: hair shaving patterns that distinguish social status (e.g., Pl. 23e; Figs. 99e; 96b,d); geometric shapes derived from ornaments worn on festival occasions (*tiltil*; Fig. 97c); special implements used in shaving or scarification (*fedeta*; Fig. 96c); cast brass knife hilts (*?inhluuta*; Fig. 99d; cf. Fig. 110a); and even the circular structure of a mud granary (*'bendewtarta*). Also included in this enumeration of distinctive cultural phenomena are the less ephemeral markers of Ga'anda status and ethnicity—scarification designs (Figs. 108,109).

Calabash decoration seems to offer a Ga'anda artist the greatest possibilities for the creation of visual statements through graphic transpositions. As suggested above, it appears that composite motifs are rarely, if ever, intended to add up to one coherently meaningful statement. Instead, the designs randomly represent various equivalents of a range of human experience, and the selections and combinations are likely to be governed by aesthetic rules and preferences. While Ga'anda artists and patrons recognize specific design equations, the *original* sources for the shapes and textures remain equivocal and are probably beside the point. This means the designs that appear to reproduce scarification motifs may have themselves evolved out of a response to other visible forms. This is suggested by the terms used for some of the

119. A comparison of *wenyiicikan'gera* triangular interlacing and the palm frond mat weave it is named after: a. Detail of Figure 54. b. Ga'anda mat. February 1980.

more dominant motifs—*caxi'yata* curves are named after circular calabash scoopers, and *kwardata* lozenges are no different in shape or orientation than the repetitive diamond-patterning of fish net.

Although the debt to the visual power and complexity of Hleeta still must be acknowledged, the weight of evidence suggests that Ga'anda design referents essentially express a keen observation of the environment and an implicit celebration of how man has processed and transformed elements of it to meet his economic, social, and cultural needs. By altering surfaces of objects with designs identifying manageable aspects of the natural and cultural world, the Ga'anda may be expressing a need to regulate less predictable social and spiritual processes. Control may be extended via design choices, as well as their execution, orientation, and juxtaposition. For example, the designs permanently and consistently scarified on women help secure their commitment to household affairs and social perpetuation. The decorated gourds and pottery used to fulfill domestic obligations both directly and symbolically reinforce this commitment to order and stability. Likewise, modeling spirit vessels according to the same aesthetic principles may give the Ga'anda a better tool for localizing, contacting, and regulating the awesome forces that ultimately dominate their cosmos. The ideological continuity between man and his spirit-benefactors, fostered by a corpus of stylistically related objects, supports the partnership believed essential to economic survival.

Despite the striking, internal consistencies in Ga'anda arts, questions still remain about the possible impact of various external factors. How has contact with neighboring groups influenced the evolution of art forms? And, how can a survey of the distribution of arts known today serve as a tool in the reconstruction of a history with few conventional resources?

These larger issues can only be addressed briefly in the following section;[25] however, a comparative analysis of decorated gourds collected among neighboring peoples contributes to an understanding of local developments and regional historical dynamics. The last chapter addresses, in very general terms, two interrelated questions posed by the distribution of decorated gourds in northeastern Nigeria: how do technical and stylistic features of similarity and dissimilarity correlate with the historical patterns deduced from other available evidence; and, how can their rich and various decorative particulars be used as a special index of more recent social interaction and change.

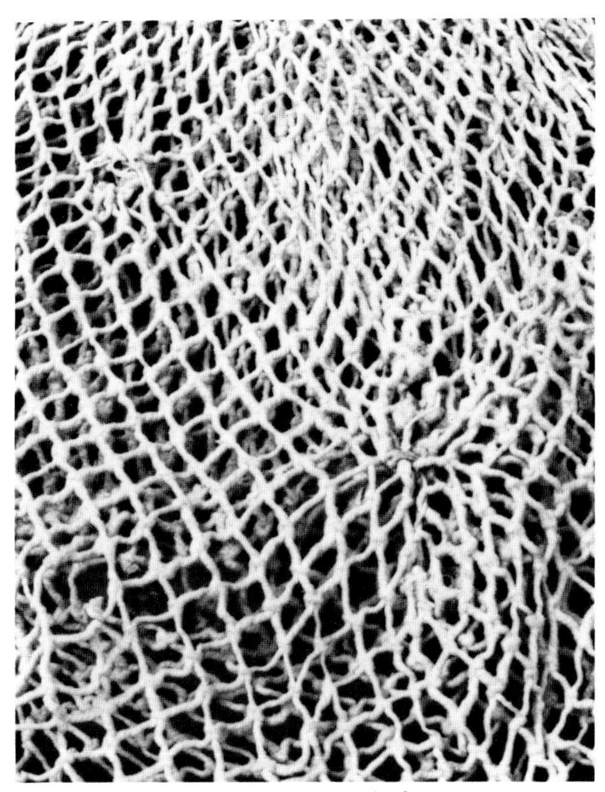

120. A comparison of *lan'taya* chained diamonds with the type of knotted fishnet they are named after:
a. Child's gourd cup (*təbwaa*). 11 cm x 11 cm. Berns collection. b. Knotted fish net. 1982.

121. A comparison of *saxta* bands with the braided fiber ropes they are named after:
a. Detail of Ga'anda porridge bowl (*tɇb'yata*). 21.8 cm x 24.3 cm. UCLA MCH X85–18. b. Braided ropes. 1981.

122. A comparison of *wanfiyca* checkerboards with the woven royal apron they are named after: a. Detail of Ga'anda water bowl (*tɇb'sayema*). UCLA MCH X85–12 (see Pl. 18e); b. Detail of cloth apron. 32.4 cm. UCLA MCH X85–113.

·5·

DECORATED GOURDS AND HISTORY

The known history of northeastern Nigeria is dominated by a succession of powerful immigrant groups—the Kanuri, the Fulani and the Hausa. Geography has strongly determined the impact of these outside forces. Peoples living in areas of open terrain were accessible to intrusive populations who introduced centralized systems of political and religious organization. Groups living in areas of higher elevation and broken terrain were more sheltered from political or cultural domination and largely remained dispersed in relatively autonomous, acephalous enclaves.

From the sixteenth century, Kanuri invaders campaigned against the unpacified peoples living south and west of the Lake Chad Basin.[1] The extent and nature of their impact are not well-understood, although it is known that the northern Bole chiefdom centered at Daniski, and later at Fika, long maintained commercial, military, and political contacts with the Kanuri.[2] Oral traditions also describe the founders of the Pabir state as members of a renegade Kanuri band who left the Lake Chad area under the direction of Yamta-ra-Wala ("Yamta the Great") in the sixteenth century.[3] According to the accounts, the powerful immigrants intermarried with the autochthonous Bura population and established the Pabir state near the modern town of Biu. They maintained extensive contacts with the Kanuri and other politically organized groups living within their sphere of influence.

The Fulani holy wars or *jihads* of the early nineteenth century begin the better known history of much of the northeast.[4] Under the leadership of Uthman dan Fodio, most of the region (except areas dominated by the Kanuri) was brought under the control of the Sokoto Caliphate. This vast conquered territory was divided into a number of emirates, five of which were situated in the northeast: Adamawa, Biu, Bauchi, Gombe, and Fika. The precolonial history of the smaller autonomous groups living in this region is one of bitter struggle against Fulani religious, political, and cultural subjugation.

The earliest records of the Fulani date to the fourteenth century, when the Arab chronicler al Makrizi noted their presence in Borno (Kirk-Greene 1969:22). During subsequent centuries, their southern migrations were motivated by the need for new pasturage rather than ideas of conquest. The cultural influence of the pastoral Fulani on the groups whose lands they used for grazing (and for whom they often served as herdsmen) is not well-understood.

The presence of both settled and pastoral Fulani in the northeast had considerable impact on developments in gourd decoration. Chappel reports that in the Adamawa area, pressure-engraving is generally associated with the pastoral Fulani and other "pagan" peoples "living in the bush" (1977:34). Whether the Fulani are responsible for introducing this decorative technique (so well-suited to a nomadic lifestyle) to the sedentary groups with whom they interacted is a question addressed below. Pyro-engraving, on the other hand, is said to reflect a more "advanced approach" (Chappel 1977:34), and in the Adamawa area it is specifically associated with the settled Fulani and other communities that adopted Islam or Christianity. Chappel argues further that the settled Fulani, who were once pastoral nomadic immigrants, introduced this technique to groups living in the areas they politically and culturally dominated (1977:34). In other instances, it is likely that pyro-engraving was adopted by groups who recognized its technical advantages, its aesthetic variability, or its association with status and progress. Nevertheless, it should not be presumed without qualification that the widespread application of this technique was strictly due to settled Fulani influence.

During the time of British control in northern Nigeria, Fulani administrators were often chosen to implement a program of indirect rule, especially in districts characterized by mixed ethnicity and marked decentralization. Settled Fulani are still the governing elite in much of the region. The Hausa have also been a dominant influence before and since British occupation. While their impact is the greatest in urban centers, the spread of Hausa language and culture has been extensive and progressive. Today, Hausa is the *lingua franca* of most northeastern peoples, with the exception of those groups who still consider speaking Fulfulde (Fulani) or Kanuri an important status determinant.[5] Itinerant Hausa artists and traders have traveled widely in northern Nigera, capitalizing on the market for a wide variety of commodities. As indicated above, specialists in gourd decoration, probably originating in metropolitan centers, are responsible for providing the pastoral Fulani with large decorated milk bowls. So widespread is this need that Hausa carvers were documented working at the Nupe market in Bida (Perani 1985:4) and at the Jukun market in Wukari for a Fulani clientele (pc: A. Rubin 1985). Though gourds decorated by the Hausa are available in many major and some minor market towns in the north, their impact on local styles and methods of working is yet to be fully explored. Hausa men use a variety of techniques, but the fact that they have long decorated their gourds by scraping or carving designs into the shell seems to counter Chappel's theory that pyro-engraving is the "advanced" technique associated with the "superior" creed of Islam. The Hausa have undergone Islamization for at least the past five centuries and were organized into powerful states long before the Fulani began their religious and political reforms (Heathcote 1976:9).

The diversity and fierce independence of the autochthonous peoples of northeastern Nigeria bring into sharp focus the problems that confronted the Fulani, as well as the British, in their attempts to impose an alien administrative structure. Even Christianity and Islam have not eroded the strong cultural traditions that give these groups their distinctive ethnic flavor.[6] Of them, gourd decoration may have a special durability because it expresses local identity and because it is an art oriented toward domestic rather than ritual or religious ends.

Although the geographic distribution of decorated gourds can effectively contribute to an historical reconstruction in the absence of written records, linguistic affiliations provide by far the most dependable evidence of relationships between groups.[7] Because the classification of languages in northeastern Nigeria is so well advanced, it serves as an especially useful guideline. Northern Nigeria is dominated by a broad belt of Chadic language speakers who in the remote past dispersed from a common base near Lake Chad (Map D). Blocks of related populations, however, have remained areally stable long enough to develop considerable differentiation, despite susceptibility to internal and external influences for at least the past 1000 years (David 1976:243–244). The two branches of Chadic that dominate the northeast—West Chadic and Biu-Mandara—suggest two directions of population dispersal: the first westward across the northern savanna and the second southward into the Mandara Mountains. People speaking languages from both branches have also migrated further west and south and have penetrated the block of Adamawa-speaking groups likely to have been aboriginal to the Lower Gongola-Upper Benue Valley.

As a result of these intrusive Chadic migrations, linguistically related peoples no longer live in contiguous zones. On the one hand, the spatial separation of groups speaking similar languages helps indicate directions of population movement and the likelihood of interaction with other historically unrelated peoples. On the other, it makes it difficult to know whether groups who now "speak the same or related languages have a common historical origin and constitute a continuation of what was a single population in the past" (Newman 1969–1970:217). Therefore, linguistic evidence needs to be cross-checked against other patterns of similarity suggested by oral traditions, ethnography, archaeology, or the arts in order to distinguish relationships due to common ancestry from those due to intergroup contact.

The gourds in this collection indicate that the greatest technical and stylistic similarities are evident between groups who have the closest geographic rather than linguistic affinity. It is also clear that artistic traditions, like languages, diversify over time but contain in their formal features information about origins and subsequent influences on them. In the historical analysis that follows, an attempt will be made to correlate styles and techniques of gourd decoration with linguistic information. To strengthen the arguments, oral traditions, ethnography, and other categories of art that have persisted over time or about which some records exist (such as pottery decoration and programs of scarification) will be synthesized. The intention here is not to detail all the categories of evidence, artistic or otherwise, that can be incorporated into a comprehensive historical reconstruction, but to reinforce the proposal made earlier that a study of one art has special value if analyzed from a distributional as well as a synthetic perspective.[8]

THE GA'ANDA HILLS

The Ga'anda Hills are the most geographically remote and isolated of all the subregions discussed here, and diverse populations were traditionally dispersed in hamlets across the terrain. The Ga'anda, speakers of a Biu-Mandara Chadic language, are an intrusive element among the predominantly Adamawa-speaking peoples of the Ga'anda Hills. The Gbinna, Yungur, and Mboi (all classified within the "Yungur group" of Adamawa) flank the Ga'anda to the west and south. This spatial distribution suggests that the Ga'anda migrated from the east, from an area of high Biu-Mandara concentration.[9] This hypothesis is supported by oral traditions collected among the Ga'anda that recount a period of occupation in the Mubi area before their ancestors moved westward and eventually settled in their present location.[10] Yungur-speaking groups, however, claim to be autochthonous and commonly trace their ancestry to a local hill site named Mukan, north of the town of Song.

A number of striking social and cultural resemblances have been noted between Yungur- and Ga'anda-speakers, despite their divergent historical origins.[11] Parallels between them suggest that intergroup contact has been longstanding and intensive.[12] Like the Ga'anda, the Yungur and the Gbinna require that girls undergo an elaborate program of facial and body scarification (Sã) and that boys complete an initiation ordeal (Xono) as prerequisites to marriage (Fig. 123). The Gbinna also build elaborate mud granaries and decorate them with motifs drawn from women's body scarification

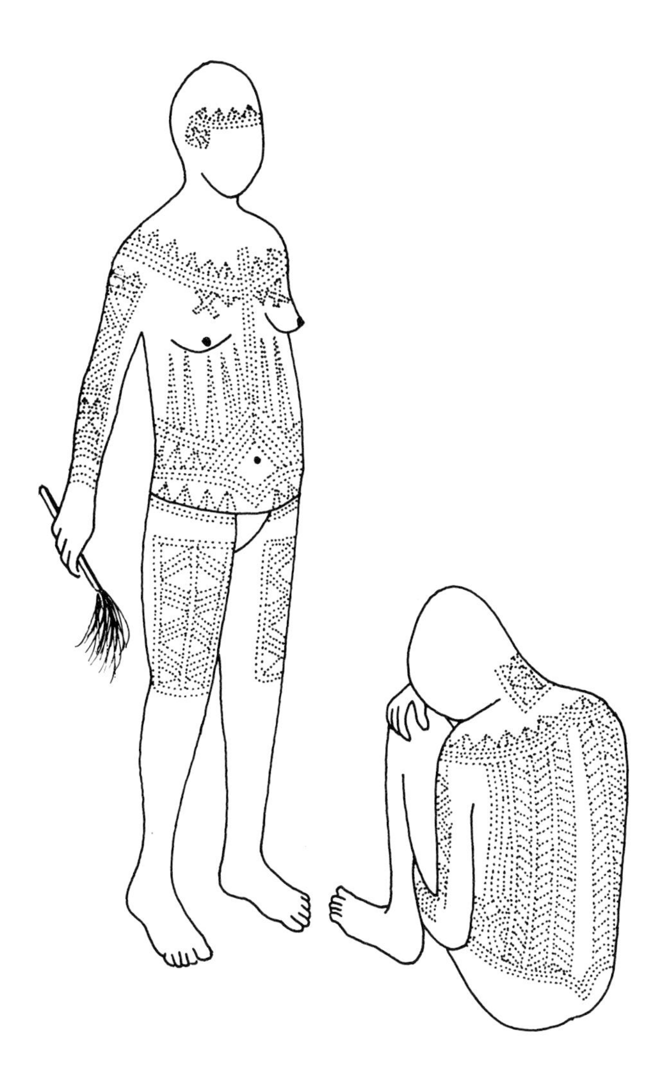

123. Gbinna facial and body scarifications (Sã); contours drawn after Chappel (1977:fig. 222).

(Fig. 124). The Yungur and Gbinna maintain shrines with elaborately modeled ceramic vessels (*wiiso*) that localize key spirit energies (Fig. 125). And, each use pressure-engraving to decorate gourds, most of which are given to girls in bridewealth payments.[13] That this decorative technique is concentrated in the Ga'anda Hills suggests its development was conditioned by the same circumstances that fostered corresponding social and cultural processes.

Despite their shared ethnographic features, the ethnic autonomy of the Ga'anda, Gbinna, and Yungur is reflected in the technical and compositional variations evident in their individual styles of pressure-engraving (Fig. 126).[14] The same stylistic individuation is evident in body scarification, ceramic sculpture, and granary construction. Nowhere, however, are the individual decorative programs of these different categories of art as integrated and consistent as among the Ga'anda.

The convergences between the disparate groups occupying the Ga'anda Hills are even more outstanding in light of the limited number of similarities between the Ga'anda and their close linguistic relatives, the Hona. The Hona now live northeast of the Ga'anda in the broken but accessible terrain of the Uba Plains. Oral traditions indicate that both groups migrated together from the Mandara Mountains to the Ga'anda Hills where they separated. While the Hona style of pressure-engraving is closely related to that of the Ga'anda and reflects their historical ties (cf. Fig. 126a,c), other aspects of ritual observance and artistic production diverge sharply. Most striking is the absence of Hona women's scarification, men's initiation, and sculpturally elaborate ceramic vessels. The geographic separation of these groups is responsible, in part, for their differences. Their dissimilarities also strengthen the likelihood that Ga'anda arts and practices were shaped by close contact with Yungur-speaking populations.

As elsewhere in the northeast, the British delegated local authority to the settled Fulani rather than to one of the many chief/priests who held power in the Ga'anda Hills. Despite their administrative control, there is little observable Fulani

124. Gbinna woman's mud granary (*pinhamo*); Riji. March 1981.

influence on Gbinna or Ga'anda culture. This may have been because no permanent Fulani settlements were ever established near their inhospitable outposts. Accordingly, aspects of traditional life have persisted more strongly here than further south in the Yungur area, more accessibly situated between the watersheds of the Gongola, Benue, and Loko rivers or further north in the Hona area. Yungur exposure and responsiveness to new ideas is responsible, at least in part, for their variable styles and techniques of gourd decoration. Informants acknowledge that pressure-engraving is the traditional carving technique (Chappel 1977:46). Pyro-engraving is an innovation taken up no earlier than the 1950s—its introduction directly attributable to the spread of settled Fulani influence. Gourds made by some Yungur women clearly illustrate the progressive adaptation of designs associated with pressure-engraving to the new technique (Chappel 1977:figs. 88–97).

Just south of the Ga'anda Hills proper live the Bata, who dominated the Upper Benue before the settled Fulani and who have embraced Islam, becoming culturally indistinguishable from the ruling Fulani elite.[15] This is especially true of Bata groups living around the administrative capital of Yola. Like the Yungur, the Bata have virtually abandoned pressure-engraving in favor of the Fulani method of pyro-engraving (Chappel 1977: 44). Pockets of Bata speakers also live further north around the town of Song near the Mboi, and gourds collected in this area exemplify a distinctive heavy engraving style and a preference for dyeing the shell a deep, rich red (Pl. 18c). Their work has influenced the decorative technique now used by the neighboring Mboi, who have also abandoned the pressure-engraving characteristic of their Ga'anda Hill relatives.[16]

Because the technique of pressure-engraving is confined primarily to this hilly zone and the nearby Longuda Plateau, the question of pastoral Fulani influence should be briefly considered. Stenning's maps of Fulani migrations and transhumant orbits (1952:figs. 3,4) and their distribution across the West African savanna (1952:fig. 1) show that the Gongola-Hawal Basin was one of the primary foci

125. Yungur ceramic ancestral vessel (*wiiso*) enshrined in the sacred precinct at Diterra. April 1981.

126. Comparison of gourds made by the Ga'anda, Gbinna, and Hona.
a. Ga'anda: 20.2 cm x 21.3 cm. UCLA MCH X85–15;
b. Gbinna: 21.3 cm x 22.5 cm. UCLA MCH X83–656;
c. Hona: 17.1 cm x 18.7 cm. UCLA MCH X85–41.

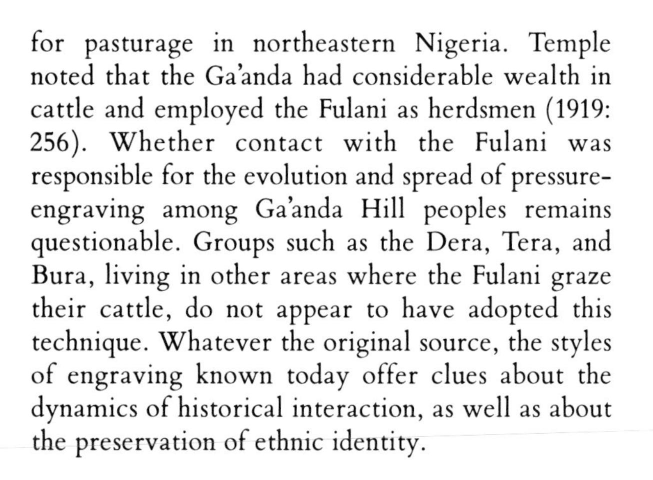

for pasturage in northeastern Nigeria. Temple noted that the Ga'anda had considerable wealth in cattle and employed the Fulani as herdsmen (1919: 256). Whether contact with the Fulani was responsible for the evolution and spread of pressure-engraving among Ga'anda Hill peoples remains questionable. Groups such as the Dera, Tera, and Bura, living in other areas where the Fulani graze their cattle, do not appear to have adopted this technique. Whatever the original source, the styles of engraving known today offer clues about the dynamics of historical interaction, as well as about the preservation of ethnic identity.

THE GONGOLA-HAWAL VALLEY

The Gongola-Hawal Valley is both ethnically and linguistically complex, largely because its fertile floodplains have long attracted migrating populations. The Dera, who live north and south of the Gongola-Hawal confluence, are organized into a network of relatively autonomous chiefdoms.[17] They speak a West Chadic language also spoken today by such geographically distant and dispersed peoples as the Bole, Ngamo, and Karekare of the Potiskum Plains, and the Tangale of the Benue Valley (Newman 1977:4).

Dera populations in the northern chiefdom of Shani and in the southern chiefdom of Shellen recount two different routes of migration.[18] The first traces the distribution of West Chadic-speaking peoples; the migration originated in Borno and passed through the Potiskum Plains before moving southward. The second follows the route taken by the "Tera group" (Tera, Jera, Ga'anda, and Hona) of Biu-Mandara; it started in the "East" and passed through Ga'anda.[19] Either way, migrant Chadic groups are likely to have encountered Adamawa-speakers, such as the Gbinna, already established in this riverain region. Thus, like the intrusive Ga'anda living to the east, the designation "Dera" may embrace a mixture of peoples with unrelated ancestries who coalesced in this fertile valley.

Northwest of the Gongola-Hawal confluence, dispersed within a forty-kilometer radius of Bima Hill, are the Tera. They speak a Biu-Mandara language closely related to Ga'anda and Hona, languages spoken over 100 kilometers away (Newman 1977:5). The territory occupied by the Tera is contiguous with that of the southern Bole so that today, the Tera are more closely related culturally to their geographical, rather than their linguistic, neighbors.

Oral traditions collected in a number of Tera villages indicate their migration took them from the Ga'anda area to Walama and then up the Gongola River to their current location. This westward movement corresponds to the present distribution of Biu-Mandara languages belonging to the "Tera group." However, other Tera villages report a joint migration with the Bole from Borno to Fika, from which the Tera continued southward down the Gongola to where they live today, a route following the direction of West Chadic (and not Biu-Mandara) population dispersal. That both traditions contain some truth (and that there were two migrations) indicates that both ancestor populations could not have been Tera-speaking.[20] This discrepancy between linguistic affiliation and origins has led Newman (1969–1970:219) to reconstruct the present Tera population as an amalgamation of peoples with Biu-Mandara and West Chadic ancestries. Such a historical phenomenon can be explained by "language-shifting"—the process whereby a group of people adopt the language of another group in place of their own (Newman 1969–1970:218).[21] For the most part, Bole-speakers have "shifted" their language to Tera. Yet contrary to expectation, present Tera art and culture do not reflect their mixed historical origins (Newman 1969–1970:220). The Tera have adopted the culture of their Bole neighbors, an observation supported by the virtual absence of ethnographic or artistic parallels between the Tera and their close linguistic relatives, the Ga'anda and the Hona.

West of the Gongola-Hawal confluence and south of the radius of Tera (and Jera) villages is the hilly zone occupied by the Waja. The Waja represent the northwesternmost extent of the belt of Adamawa speakers dominating the Lower Gongola Valley. Their geographic proximity to the Tera has left a distinctive Chadic mark on their arts and culture.

The Bura dominate the elevated grasslands of the Biu Plateau northeast of the Gongola-Hawal confluence. Like their western Tera neighbors, among whom some of them live, they speak a Biu-Mandara Chadic language. However, Bura is classified within a different linguistic subgroup, linking them with populations further east in the Uba Plains—the Margi, Kilba, and Chibak (Newman 1977:5–6). The Bura represent the western frontier of a migratory wave described by the spatial distribution of these related groups. It probably moved out of the Mandara Mountains at a more northerly latitude than the Ga'anda-Tera wave that penetrated the hills southeast of the Hawal River. The Bura speak the same language as their northern neighbors, the Pabir, who have a long history of centralized political organization and adherence to Islam. Bura villages, however, remained culturally isolated and politically autonomous until recently.

Today, the residents of most Gongola-Hawal Valley towns have adopted either Islam or Christianity and have abandoned many of their traditional ways of life. Yet, what is known about their arts and practices reflects their ethnohistorical complexity. This is particularly true of the Dera, who also have historical ties with groups living in the Ga'anda Hills. Early reports indicate that Dera architectural decoration, boys' initiations, and girls' scarification relate to practices still maintained by their eastern neighbors (the Ga'anda, Gbinna, and Yungur).[22] The Dera of Shani maintained close ritual ties with the Ga'anda, whose control over rainfall extended across the forty-kilometer distance between them.

Despite such parallels, the Dera decorate their gourds with pyro-engraved designs, not with the pressure-engraving typical of their eastern neighbors. The Dera engraving style is closely related to the distinctive painted ornamentation found on their display pottery. This formal relationship between gourd and pottery decoration links the Dera to the other groups living in the Gongola-Hawal Valley—the Tera, Bole, Bura, and Waja.

Like the Tera, Dera women encircle the inner perimeters of their rooms with displays of pottery and calabashes. The most distinctive ceramic containers (*denga puro*, "pots for [standing in] sand"), also used during public festivals for holding beer, are built with complex, segmented profiles and sometimes intricate openwork handles (Fig. 127). They are decorated with white, painted designs and bands of wood rouletting. The same elaborately painted patterns are found on smaller containers (*wunda*) which, like decorated gourds, are carried on young women's heads during Menwara dancing (Figs. 128,29).[23] These geometric motifs and fine linear patterns are very similar to the designs engraved on Dera gourds—indeed, the "burned in" lines reverse the effect of white paint on a redware ground (Fig. 129). It is possible that the Dera custom of abrading the gourd's cuticle to accentuate the contrast between figure and ground was inspired by the color contrasts typical of their painted pottery. A comparison of Figures 127 and 128 with Figure 129 reveals the curvilinearity and complex textures that dominate both design compositions. Even the openwork handles that distin-

127. Pidlimndi painted display vessel (*bangelan'~~de~~maham*); Walama.
Ceramic and pigments. December 1981.

128 Dera painted dance vessel (*wunda*);
Kurgulum (Gasi subgroup). Ceramic and pigments.
36.2 cm. LX 83–50. Promised gift of Jeanne and Jim Pieper.

129. Dera pyro-engraved gourds (*lib'e muni*).
a. 20.9 cm x 23.5 cm.
UCLA MCH X83–677;
b. 21.6 cm x 23.2 cm.
UCLA MCH X83–669;
c. 16.5 cm x 17.7 cm.
UCLA MCH X83–667;
d. 19.7 cm x 21.6 cm.
UCLA MCH X83–662.

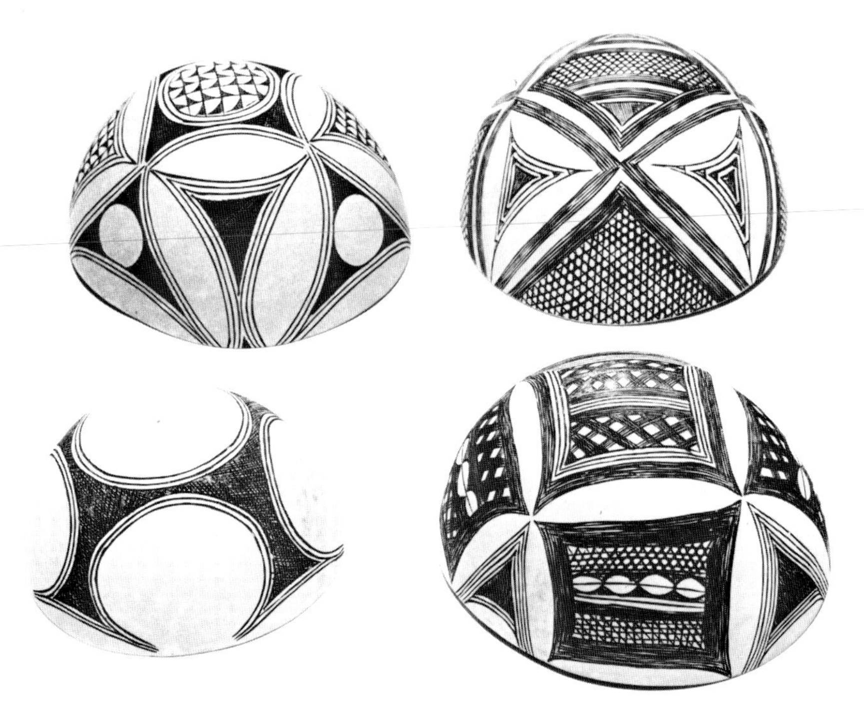

guish many larger *denga puro* and smaller *wunda* have distinctly curved contours and circular piercings.

On ethnological grounds, the Pidlimndi, who speak a Tera dialect (Biu-Mandara), are indistinguishable from their Dera (West Chadic) neighbors (Newman 1969–1970:221). They decorate their gourds in the same ways (Pl. 22; Fig. 75) and make identical display vessels (*bangelan'đemaham*) for the same purposes (Fig. 127). Their social and cultural assimilation is explained by language shifting, in this case a Dera population settling in the village of Walama and later adopting the language of the Pidlimndi living there (Newman 1969–1970:221). Their present situation reflects the political and cultural dominance of the Dera in this area, but retention of the Pidlimndi language infers their ancestors had historical ties with the Tera.

Further north in the Gongola-Hawal Valley, close contact between the Tera and their eastern Bura neighbors is clearly reflected in their arts and culture, as demonstrated above with respect to their shared styles of gourd decoration.[24] Painted pottery also appears to have offered the Tera and Bura a source of inspiration for the evolution of gourd designs. Their pottery decoration was influenced by the distinctive painted vessels made by the Bole.[25]

130. Tera painted ceramic vessel for stacking gourds (*kayanghladi*). 48 cm. April 1982.

Early colonial reports note the rows of pots stacked one on top of another arranged around Bole rooms (Temple 1919:68). This convention is the same as that described and illustrated earlier for the Tera, who probably adopted it from the Bole (Pl. 9). Bole vessels with "hourglass" contours are modeled expressly for stands for room displays.[26] Like those made by the Dera, they are covered with registers of linear designs painted in white. Display vessels made by the Tera also are elaborately painted (*kayanghladi*; Fig. 130). In this case, the arrangement of motifs in a series of encircling bands and/or vertical registers is very like the orientation of engraved designs on Tera gourds (Fig. 131). The pot illustrated here is painted with short feathery or dotted lines that recall a characteristic feature of Tera engraving—*ndesa* (Fig. 77).

131. Tera pyro-engraved gourd (*d'eba*; interior is Fig. 51b). Berns collection.

Tera water pots (*shongom*) also have painted designs that resemble the motifs engraved on gourds (cf. Figs. 132,133). Vessels built with identical contours and decorated with similar zones of painted patterns were collected in Fika early in the century, suggesting that in this case as well, the Tera *shongom* is based on a Bole model (see Hambly 1935:pl. XCV). The Bura also make water pots with similarly segmented contours (*bangilang*) and

132. Tera water pot (*shongom*). 48 cm. Ceramic and pigments. April 1982.

133. Jera pyro-engraved gourd (*d'eba*). 28.2 cm x 28.2 cm. UCLA MCH X83–707.

decorate them with designs painted in white (see Leith-Ross 1970:42; no. 220). Though more restrained in character, the incised and painted decoration recalls the densely engraved style of some Bura gourds.

Pots built with elongated segmented necks are widely distributed among Chadic groups related to one another either by language or by geography—the Ga'anda, Dera, Tera, Bole, and Bura. Nevertheless, today, the way the Ga'anda impress a series of distinctive and elaborate textures into the surface of their vessels differs considerably from the way other groups paint fine linear designs. It may be noted that pressure-engraving in gourd decoration seems to correlate closely to the meticulous incision of dense textures on pottery, while pyro-engraving seems to relate to careful, but more loosely applied painted motifs. Considered historically, these divergent technical orientations may reinforce the distinctions between Biu-Mandara and West Chadic sources.

The Tera, Jera, southern Bole, and Bura also share the practice of marking the faces of infants with long vertical striations (Figs. 134, 135). Using a triangular razor, cuts are made on the forehead, down each cheek, and/or along the chin. A deep central incision extending from the forehead to the tip of the nose is the first to be made and serves to organize the symmetrical disposition of lines on either side of the face. However, the specific alignment, depth, density, and number of cuts differ from group to group, and even within each group. These distinctions, though sometimes subtle, are easily recognized by the people themselves and show that ethnic autonomy prevails.

The northern Bole of Fika, as well as the Ngamo, Karekare, and Ngizim living around Potiskum, have similar programs of facial scarification (Fig. 135b-e).[27] As all these groups speak related West Chadic languages, it is likely that the Bole introduced this mode of ethnic identification to the Gongola-Hawal Valley. Moreover, the Waja and two other related Adamawa groups—the Awak and Kamo—have the same traditions of facial scarification. The markings on these groups are among the most complex and remarkable of any observed in this region (Fig. 136). Long vertical striations and patterns of short diagonal lines are also incised down the chests and abdomens of women. This tradition of scarification, as well as styles of gourd decoration and pottery display, confirms the proposition that the Waja (and other geographically proximate groups) have been strongly influenced and perhaps politically dominated by their northern Tera, Jera, and Bole neighbors (Fig. 137).

134. Facial markings on Jumai Pitiri Gulcoss, a well-known Tera artist; Wuyo. 1969.

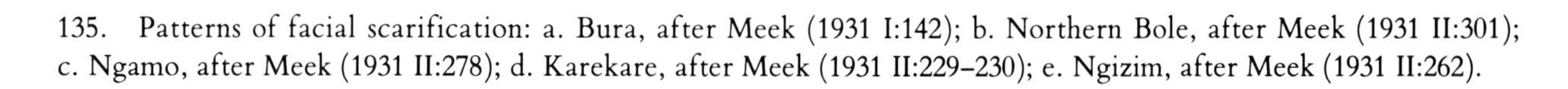

135. Patterns of facial scarification: a. Bura, after Meek (1931 I:142); b. Northern Bole, after Meek (1931 II:301); c. Ngamo, after Meek (1931 II:278); d. Karekare, after Meek (1931 II:229–230); e. Ngizim, after Meek (1931 II:262).

136. Facial markings (*datah*) on the Chief of Kamo's wife. March 1982.

137. Waja ceramic vessels (*jalum*) stacked with decorated gourds (*bɐla kaage*) as they would be displayed inside a woman's room; Swa. February 1982.

The Kanuri, who consolidated considerable power in the Lake Chad Basin by the ninth to tenth centuries A.D., have a similar pattern of vertical facial scars (Fig. 89).[28] There is little evidence to indicate how Kanuri domination of Borno affected the Chadic groups living there—either through assimilation or by expulsion—or whether the traditions long maintained by Chadic groups were adopted by their Kanuri invaders. The Bole argue that they migrated into the Borno area along with the Kanuri, a claim implausible on linguistic grounds but one that associates the Bole with a group that occupies a powerful place in Nigerian history.[29] Regardless of its original source, this ubiquitous mode of scarification, likely to have been practiced by peoples for many centuries, provides an important clue about the dynamics of historical interaction in northeastern Nigeria.[30]

The technique used by the Tera, Jera, Bura, and Waja to incise continuous grooves on the face is different from that used by the Ga'anda, Gbinna, and Yungur to achieve rows of small, dense nodular cicatrices. This difference, like modes of pottery decoration, relates to the contrast between burning designs into the surface of a gourd with a hot knife and pulling lines across the suface using an iron point. As suggested earlier, the laborious method of pressure-engraving used by the Ga'anda Hill groups requires the same control and multiple movements as does their technique of scarification. In parallel fashion, the same groups who use the quicker and more sweeping movements of pyro-engraving also incise broader facial striations. Among the Tera and Waja, the predominance of feathery or closely hatched lines on calabashes corresponds to the close parallel lines of scarification, and informants often name engraved designs accordingly (Fig. 138). Moreover, triangular motifs named after the shape of the razor used to incise the striations—*sud'exas*—are commonly found on Tera gourds (see Figs. 131, 133). The fact that the Tera, Bura, and Waja use similar techniques and designs to decorate gourds, pots, and their faces strengthens the linguistic argument that these composite peoples converged in the fertile plains and scattered hills of the Gongola-Hawal Valley. Likewise, pressure-engraved gourds and nodular cicatrices are mostly found among the peoples who live in the craggy hills east of the Gongola.

Adding to the ethnic complexity of the Gongola-Hawal Valley are the Fulani. Pastoral groups have long grazed large herds of cattle along the fertile flood plains, and settlements of semisedentary Fulani clans were established as early as the mid-1700s.[31] Low proposed that the latter group of Fulani, most of whom had not yet adopted Islam, were likely to have taken up the faiths and cultures of their more settled neighbors, which would have included the Dera, Tera, Jera, Waja, Pabir, and Pidlimndi (1972:86). He does not mention,

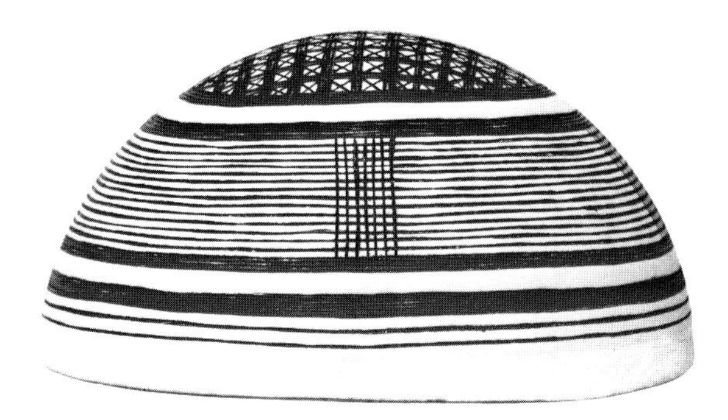

138. Waja gourds with motifs named after patterns of facial scarification:
a. *Jayya* vertical striations. 17.8 cm x 20.3 cm. UCLA MCH X85–50;
b. *Daula* diagonal strokes. 23.6 cm x 24.6 cm. UCLA MCH X85–51.

however, whether these intrusive Fulani clans culturally impacted the indigenous populations among whom they lived. It is possible the tradition of decorating calabashes, so much a part of the Fulani way of life, presented a source of artistic inspiration to local groups. The clearest evidence that this happened in the Gongola Valley is reflected in the style (but not the technique) of decoration used by the Tera, Jera, and Waja (Pls. 24,30). The vertical orientation of designs around the rim of the gourd, like spokes converging on the center, is a likely adaptation of one pastoral Fulani style of pressure-engraving (Fig. 69). That all these groups identify some vertical motifs with patterns of facial scarification further reinforces this suggestion. In addition, the way gourds are displayed by the agrarian Tera to enhance a woman's status may be adapted from pastoral Fulani practice.

POTISKUM PLAINS

Unlike the first two subregions, the Potiskum Plains are linguistically and geographically homogeneous. The groups around the towns of Potiskum and Fika—the Bole, Karekare, Ngamo, and Ngizim—all speak West Chadic languages. These languages are classified together within the "Bole group" of West Chadic's subbranch "A," indicating that except for the Ngizim, geographic affiliations correspond with linguistic relatedness (Newman 1977:4). Ngizim is a subbranch "B" language and the spatial separation of its speakers from other "Bole group" speakers suggests that the Ngizim are intrusive to the Potiskum area.[32]

Information available on the area indicates that the Bole of Fika (politically centralized and Islamized for several hundred years) are very different from the nominally independent "pagan" communities around them.[33] Despite their organization and military links to other powerful northern states, Bole suzerainty over the Ngamo and sections of the Ngizim and Karekare was mostly limited to receiving an annual tribute of cloth (Meek 1931 II:289). Only after colonial reorganization were the Bole made the indirect rulers of Potiskum, which became an important administrative center in the northern savanna.

According to Temple, the Ngizim moved southward and conquered the Karekare around Potiskum in 1790 (1919:310). While this date cannot be confirmed, the direction of migration correlates with what can be inferred from linguistic geography. Additionally, the chronicle of the reign of Mai Idris Alooma, who ruled Borno from ca. 1569 to 1600, details the vigorous Kanuri campaigns against the Ngizim (Palmer 1926:12–13). While Mai Idris had only partial success in breaking their fierce resistance, the relentless militarism of the Kanuri may have finally driven the Ngizim into the Potiskum area. None of the Potiskum groups held a position of centralized authority, and all remained dispersed in a number of acephalous enclaves. The Karekare and the Ngamo, who flank the Ngizim to the south and west, traditionally built compounds at the top of cliffs, which supports the proposition that defensible positions were necessary in this open and accessible terrain (Temple 1919:110,225).

Despite differences in social, political, and religious organization the historical relationship of Potiskum peoples is suggested not only by their closely related languages, but also by shared cultural traditions. Each group is distinguished by a particular pattern of facial striations (Figs. 134–136), and the correspondences between them suggest this practice derived from a common source. Similarly, Karekare, Ngamo, and Ngizim women all paint the interiors of their gourds with elaborate designs (Fig. 139; see Pls. 20, 32). As indicated earlier, the Karekare claim to be at the center of this tradition due to their control over the mining and processing of the clays used as pigments. Whether or not the Karekare developed and then introduced this practice to their neighbors is a question that cannot be answered on the basis of available information.[34] What can be considered, however, are the historical implications of parallels between this method of decorating gourds and the way related West Chadic groups paint their pottery. Both types of highly decorated household objects are made to be displayed in women's rooms. On pottery and on gourds, fine white and/or black linear patterns are painted over a red ground. The ceramic jugs and pot-stands made by the Bole and the Tera are painted with zones of linear motifs whose spacing and alignment resemble the interior decoration of Ngamo and Ngizim gourds. Fine linear cross-hatching and other specific motifs also relate to the painted designs on Dera vessels (Figs. 127,128). While the examples chosen to illustrate this point have specific stylistic affinities, what is most striking about the artistic traditions maintained by West Chadic (or West Chadic-related) groups is the primary use of pigment to create a decorative overlay and its application in dense repeated patterns.

The ubiquity of this decorative tradition strengthens the proposition that the compositions and zones

139. Ngizim gourd with painted interior. 28.9 cm x 29.8 cm. UCLA MCH X83–772.

of texture that characterize Tera, Bura, and Dera gourds are adaptations of painted designs to a new pyro-engraving technique. It is possible that, in the Gongola-Hawal Valley, pigments are only used in pottery decoration because pyro-engraving is a more direct and flexible decorative alternative. It eliminates the need for undertaking the complicated process of producing multicolored pigments from clay. However, this proposed technical shift also means that designs once used in a program of interior gourd decoration were translated into designs now concentrated on the exterior. It may be for this reason that the same groups who adopted pyro-engraving also favor the rather distinctive and unusual practice of engraving the insides of gourds with related designs (Fig. 140a,b).

The Potiskum area lies along a transhumant route of the pastoral Fulani (Stenning 1952:fig. 1). As mentioned above, pressure-engraved gourds are sold by the Fulani to local women who paint their interiors using conventional designs. Some Ngizim, Ngamo, and Karekare women have learned the Fulani technique and use it to decorate the exteriors of their own gourds.

Northern Bole men decorate gourds both for a local clientele and to sell at large markets.[35] This practice implies the influence of Hausa culture, where male artists and traders produce and distribute such household items. While the influx of Hausa residents into this region is fairly recent, the location of Potiskum between Maiduguri and Kano has long made it accessible to traders coming from either of these major market towns.

UBA PLAINS

The Uba Plains can also be characterized as a linguistically homogeneous area. The Margi, Kilba, and Chibak all speak Biu-Mandara Chadic languages classified within the "Bura" subgroup (Newman 1977:6). In this division, the western Margi (Putai) and the Chibak are more closely related to the Bura, whereas the eastern and southern Margi are more closely related to the Kilba. The distribution of these peoples from the western slopes of the Mandara Mountains across the Uba Plains to the Biu Plateau implies their ancestors migrated westward and southwestward. Where each group settled, of course, conditioned their subsequent culture history. The western location of the Bura accounts for their cultural affinities with the Tera, which are evident in gourd decoration and facial scarification. The Chibak and the Putai live in the open terrain around Askira and have undergone considerable cultural integration.

The correlation between relative geographic proximity and linguistic relatedness implies that the Margi, Kilba, and Chibak represent a continuation of what was a single population in the past. To support this inference, Meek notes that these groups share certain ethnographic features: the practice of initiation rites, the removal of the epidermis from the bodies of deceased elders, rules of exogamy, and systems of centralized chieftainship (1931 I:214). On balance, however, each group has maintained a high degree of local autonomy, a factor even more remarkable in light of the intrusive groups who moved into or through this open area southwest of Lake Chad. According to Meek, both the Pabir and the Kanuri settled the area and in some cases coalesced with local populations (1931 I:214–216). This influence is probably reflected in the facial markings worn by Margi and Kilba women—a number of vertical cuts generally flanking a central striation running from the tip of the nose to the forehead (see Fig. 66; Meek 1931 I:233). Yet, at the same time, both the Kilba and Margi maintain traditions of body cicatrization, linking them to their Biu-Mandara speaking neighbors to the east in the Mandara Mountains, as well as to the west in the Ga'anda Hills.[36]

In the nineteenth century, groups of settled Fulani established communities around the Margi, Kilba, and Chibak. Some Fulani moved southward out of Borno and others extended the authority of the Adamawa emirate centered at Yola. The pyro-engraved gourds collected among the peoples of the Uba Plains resemble those made by the settled Fulani both in technique and in design composition. However, of these groups, neither the Kilba nor the eastern Margi were ever conquered by the Fulani because of their relative inaccessibility and because their centralized political organization gave them a better basis for resisting subjugation (Vaughan 1981:96). Even after the colonial government placed these groups under Fulani authority, they retained a considerable degree of local autonomy. It is noteworthy that despite efforts to maintain their ethnic identity in political and cultural terms, such groups did not find it compromising to adopt a decorative technique so closely linked to their Fulani oppressors.[37] This may have been because neither the presence of gourds in daily life nor the activity of decorating them was new. While not embracing the religion or way of life of the Fulani, this open borrowing of a decorative technique was a modest way of expressing modernity and political elitism. As for Gongola-Hawal groups, the ability to burn designs directly

140. Dera pyro-engraved gourd (*lib'e tibe'kokumɇt*).
a. Interior; b. Exterior. 17 cm x 19.2 cm.
Berns collection.

into the shell of the gourd was probably an attractive technical alternative to other more time-consuming procedures. The ease with which stylistic variations can be achieved through pyro-engraving may have provided these groups with a vehicle for restating their ethnic individuality in artistic terms. This is especially clear in the differences between Kilba, Chibak, and Margi gourds, despite compositional features that reveal their shared Fulani models. The multigauged linear hatchings typical of the Chibak, Putai (western Margi), and Kilba reflect the particularly close relationship of these groups, especially in the Askira area (Pl. 28a; Figs. 48,141).

The way motifs are arranged around the rim of Chibak (and Putai) gourds also suggests that a different method of decoration may have been adapted to pyro-engraving. Askira is situated along a migratory route of the pastoral Fulani who herded cattle belonging to indigenous groups living in the adjacent plains, as did other Fulani for the Ga'anda (Meek 1931 I:181). It is possible that the Chibak incorporated design features of pastoral Fulani pressure-engraving (Fig. 80), particularly the way large geometric shapes create a quadrapartite division of the design field (Fig. 141). Accordingly, the broad-groove alternations, typical of the pastoral Fulani engraving style may have inspired the linear cross-hatching of the Chibak (and even the settled Fulani). It is harder to determine the influence of the pastoral Fulani on groups with whom they long have had economic transactions than it is to see the more recent (and often acknowledged) impact of their settled brethren. The lower status accorded these nomadic people has probably discouraged sedentary groups from acknowledging cultural borrowings, such as modes of decorating or using gourds. In contrast, the high political and cultural status of the settled Fulani were admittedly positive determinants in the adoption of pyro-engraving.

The eastern Margi live in a protected area along the western slopes of the Mandara Mountains, an area marked by pronounced ethnic heterogeneity. The Margi are organized into divine kingdoms and share a number of social and cultural traits with their immediate Biu-Mandara neighbors. The most distinctive of these characteristics is a caste system that separates an endogamous group of blacksmiths from the rest of the population.[38] Although the Kilba and Bura do not have a caste system, they are known for their iron smelting and smithing skills. All three groups use iron amulets as the focus of religious worship, iron spears as emblems

141. Chibak pyro-engraved gourd. 34.9 cm x 37.4 cm. UCLA MCH X83–770.

of leadership, and iron ornaments in programs of self-decoration. While in other areas ceramic sculpture seems to be a primary artistic device for suggesting historical relationships, among these groups the distribution of similar forms of iron regalia is an indicator of common origins and intergroup contact.[39]

Among the Uba Plains groups, only the eastern Margi carve designs into the outer cuticle of gourds and then resist-dye them red. This is analogous to the Ga'anda technique of pressure-engraving the design, reversing it with charcoal and oil, and then dyeing the gourds red. That these related methods of decoration are confined to groups living in relatively inaccessible hilly areas supports Chappel's contention that they are older than pyro-engraving in the northeast (i.e., they predate the nineteenth-century domination of the settled Fulani; 1977:34). Some elaborately pressure-engraved or resist-dyed gourds have been documented among groups living in the mountains east of the Uba Plains along the Nigerian border.[40] Sometimes minimal pyro-engraving is also used, and as elsewhere in the Uba Plains, this technique is likely to have been adopted by Mandara Mountain peoples from Fulani examples (Wente-Lukas 1977a:41). The Margi, who also use pyrogravure, admit to imitating the work of others and, as suggested above identify one of the most popular designs as "Fulani" (Vaughan 1975:185).

BENUE-GONGOLA VALLEY

Like the Gongola-Hawal confluence area, the Benue-Gongola Valley is ethnically and linguistically complex, having long attracted migrating populations. Today, the plains flanking the Upper Benue are dominated by the Chadic-speaking Bata, their Bachama relatives, and a substantial population of settled Fulani. That these groups are intrusive is suggested by the degree of linguistic differentiation evident in the blocks of Adamawa-speaking peoples living both north and south of the river (where they are concentrated in the greatest numbers).[41]

The Mbula live just northeast of the Benue-Gongola confluence and speak one of the Jarawan Bantu languages found today among peoples living in a wide geographic area extending from southeastern Cameroon to the Jos Plateau (Map D). The most linguistically homogeneous Jarawan Bantu area is located west of the Benue-Gongola confluence in the hills south of Bauchi. The idiosyncratic dialect spoken by the Mbula suggests that, instead of migrating westward with their relatives, they remained an isolated enclave (Ballard 1971:299).

Oral traditions confirm the historical inferences drawn from the linguistic geography of the Benue-Gongola Valley. Most Adamawa-speaking groups recount stories of origin that name local hill sites as sacred homelands, claims that support their aboriginal status in the region.[42] The Chamba, an Adamawa group, probably dominated the Upper Benue Valley until they were displaced southward and westward in the eighteenth century by a wave of Chadic-speaking Bata migrants from the northeast.[43] Thereafter, the Bata controlled both banks of the Benue from Yola to Garoua (northern Cameroon) until the nineteenth-century conquest by the settled Fulani. Probably before this period of Fulani expansion, the Bachama separated from their Bata brethren following a chieftaincy dispute (Stevens 1976:31). The Mbula trace their origins to Rei Buba, an area in eastern Cameroon that accords with the linguistic geography of Jarawan Bantu peoples. Curiously, another Mbula account claims that they traveled down the Benue along with the Bata (Kirk-Greene 1969:169). It is more likely that the Bata entered the Upper Benue region after the Mbula were established, making it questionable that the Mbula settled "with the permission of the Bata" (Kirk-Greene 1969:170; Nissen 1968:129). Other stories describing how the Mbula were driven southwestward by the advance of their northern Yungur neighbors support the possibility of their prior settlement in the region (Kirk-Greene 1969:170).

Information about Mbula ethnography confirms that they had close historical contact with their linguistically disparate neighbors—the Bata, Bachama, and Yungur. For example, Meek draws parallels between the Mbula and the Bata, especially regarding social organization, burial procedures, and aspects of religious worship (1931 I:62). The Mbula, Bata, and Bachama also share the powerful cult of Nzeanzo, the deity of health and prosperity.[44] Other Mbula religious practices resemble those of their Yungur neighbors; they produce large figurated ceramic vessels as receptacles for spirits of deceased chiefs from whom they solicit assistance through libations. In spite of this evidence of historical interaction and possible amalgamation, the Mbula have maintained their political and cultural autonomy. They have not, as Meek projected, lost their "tribal identity" in the course of two generations (1931 I:57). The descriptions of Mbula secret societies and burial displays

142. Longuda ceramic healing vessels (*kwandalha*); Dangir. November 1981.

provided above confirm the persistence of local traditions.[45]

The style of pyro-engraving used by Mbula women reflects an effort to maintain their distinct identity (Fig. 79). This decorative technique was probably introduced to the Mbula by the settled Fulani either directly or via their Bata neighbors. The gourds engraved by the Mbula closely resemble those Chappel classified as a separate Bata substyle (1977:44–46; figs. 73–79). Their stylistic relationship is explained by the provenance of Chappel's Bata gourds—all come from Geren, a village situated just east of the main Mbula town of Borrong. Gourds engraved by the Bachama also display some of the same stylistic and technical characteristics as those made by the Mbula, i.e., bands of heavy scorching and densely incised lines.[46] Despite the way gourd decoration reinforces the pattern of cultural interaction evident between these three neighboring groups, the asymmetrical compositions on Mbula gourds reflect a unique aesthetic response.

The congeries of small Adamawa-speaking groups in the hills northwest of the Benue-Gongola confluence—the Mwona, Dadiya, and Tula among others—do not have traditions of gourd decoration comparable to those maintained by their eastern and northern neighbors. For example, the pyro-engraved gourds used by the Dadiya to make dance rattles (*kichibyok*; Fig. 26) are also used in domestic contexts. However, their designs are very similar to those used by their Bachama and Mbula neighbors. The most distinctive tradition shared by all these groups is the incorporation of highly expressive figurated ceramic vessels in ritual contexts. Pots used for identifying and pacifying disease-causing spirits are most common, and they conform stylistically and functionally to those made by their Adamawa-speaking relatives—Longuda, Gbinna, Yungur, and Waja (Fig. 142; see also Fig. 125). Even though the geographic location of these linguistically related groups has placed them in contact with Chadic peoples, the concrete evidence that survives (especially the vessels discarded in abandoned hill sites or shrines) suggests that these ceramic traditions are a "core" Adamawa feature.[47]

The Longuda occupy the plains on the right bank of the Gongola opposite the Dera town of Shellen and the hills of the adjacent plateau. The rain priest of Dukul, a village high on the plateau, has traditionally exercised considerable spiritual and moral authority over the group (Kirk-Greene 1969:173). In the nineteenth century, Longuda warriors descended to the plains and drove the Dera from Lakumna across the river to Shellen (Meek 1931 II:333). The Longuda then founded a number of villages in the plains, of which Guyuk has become the main administrative center. The chief of Guyuk acts as the secular "mouthpiece" of the Dukul priest among the plains population (Kirk-Greene 1969:173).

Although both the occupants of the plains and the hills regard themselves as Longuda, there are marked cultural differences between them. Unlike the series of dispersed and autonomous hill communities, the Plains Longuda adopted the principles of Dera political organization, which gave the Guyuk chief authority over all villages established along the river. In due course, the Plains Longuda adopted the dress of the Dera, as well as their method of building houses (Meek 1931 II:333). Dera influence is also evident in calabash and ceramic decoration, a borrowing explained in part by intermarriage between these two peoples. In 1982, all the clay vessels exhibited by one wife of the Guyuk chief were like the painted wares described earlier for the Dera; similarly, all her calabashes had been pyro-engraved in a Dera fashion. Despite expansion and modernization in Guyuk over the past ten years, there has been little or no change in modes of calabash or pottery decoration among Longuda hill communities. They still pressure-engrave with a distinctive all-over pattern of small bisected ovals (Fig. 70), and ceramic vessels are covered with comparably elaborate rouletted designs. Because of admiration for the Shellen dynasty, adopting Dera material culture was clearly a way for the Plains Longuda to enhance their own political and cultural prestige. As elsewhere, pressure-engraving is associated with traditional "pagan" hill communities, reinforcing the proposition that pyro-engraving is a manifestation of change.

·6·

GOURDS AND MODERN CHANGE

Tradition vs. New Directions

The impressive variety of decorated gourds from northeastern Nigeria reflects the dynamic and complex history of its peoples. Today, pressures to conform to a synthetic national culture have accelerated change, and have challenged the future viability of this traditional mode of artistic expression. Two interlocking factors have shaped gourd decoration over the last two decades: the increasing popularity of the enamel, glass, plastic, and metal containers flooding major and minor Nigerian markets (Pl. 37); and the progressive abandonment of the social, ceremonial, and religious contexts in which gourds were both useful and expressive elements. Participation in Nigerian national culture varies, however, from region to region, as well as from group to group. Now, as in the past, the vulnerability or responsiveness to intrusive forces of change are conditioned by differences in geographic situation and in political and/or religious incorporation. Additionally, the sharp contrasts between urban and rural life are reflected in the ways and degrees this art form has persisted in modern Nigeria.

Until the late 1960s and early 1970s, decorated gourds were highly visible in both town and village settings. In most areas, surplus goods were transported to market and then sold in decorated carriers (Frontispiece; Pl. 6; Fig. 12). Tera, Bura, and Waja women protected babies' heads from sun and rain with pyro-engraved gourd bonnets (Pl. 10; Fig. 23). And, everywhere, food, water, and beer were served in calabash bowls of many sizes, ornamented inside or out.

By the early 1980s, the incursion of manufactured items—tinware bowls and trays, glass bottles, and plasticware—had become apparent. Particularly in rural areas, the decreased use of decorated gourds in the panorama of daily life was especially noticeable. Now many more women use large enamel basins when they go to farm or market, a change especially dramatic among pastoral Fulani groups. Power-driven grain mills, ubiquitous even in remote areas, attract a daily stream of women and girls bearing tin basins of guinea corn or maize. Food is often cooked in aluminum pots and served in lidded enamel containers, glass casseroles, or earthenware bowls; soups or stews are ladled with aluminum skeuomorphs of gourd spoons (Fig. 143). Moreover, people use plastic mugs as dippers for water, metal tablespoons for eating rice, thermos flasks for holding water, china cups and saucers for hot drinks, glass tumblers for cold drinks, and glass bottles and jars for oils, ointments, and spices.

There are a number of reasons why factory-produced household objects are so widely favored: they are more durable, they can be cleaned easily, and they are not prone to insect infestation or mold. Additionally, their brightly colored surfaces and long-lasting stencil designs are aesthetically appealing. Furthermore, they reflect the high social standing associated with a contemporary, urban way of life, serving as modern attributes of wealth and prosperity. There is little doubt that an urban context fosters new measures of social and economic status requiring the ownership of prestige wares. Gourd bowls are associated with a rural and more "backward" way of life. Women living

143. Aluminum skeuomorphs of gourd spoons.
a. 19 cm x 6.6 cm. UCLA MCH X83–809; b. 18.7 cm x 6 cm. UCLA MCH X83–808.

in outlying areas, striving to keep in step with contemporary trends, are thus eager collectors and users of modern manufactured household equipment.

The importance of accumulating capital and exchanging it for emblems of modernity and cosmopolitan status is clearly pervasive in the towns and provincial villages of northeastern Nigeria. Among the Hausa, displays of astonishing numbers of brightly colored enamelware containers have almost entirely superceded the formerly lavish exhibits of decorated calabashes (Heathcote 1976: 46). In Ribadu, a large semirural Bole village, the same substitution had been made for a traditional display of pottery and calabashes (cf. Figs. 144,145). As mentioned earlier, even in remote villages in the Ga'anda Hills, the trousseau of a new bride can consist entirely of factory-made household items.[1] At every level of organization, therefore, a woman's wealth and status continues to be determined by the number and quality of domestic artifacts she acquires, although the nature of these items has changed drastically. Also, in traditional ceremonial contexts, such as a Yungur second funeral, gifts of food and homespun cloth are more impressive when given in enamel basins than in decorated gourd bowls (Fig. 146). The substitution of factory-made containers in these contexts signals the erosion of the social and ceremonial bases for investing the considerable amount of time, labor, imagination, and capital in gourd production and decoration. The progressive conversion of communities to Islam and Christianity is also leading to the abandonment of traditional belief systems and the objects, such as gourds, that were instrumental to their efficacy.[2]

During two years of fieldwork in the Lower Gongola Valley (1980–1982), Berns was never once offered food in anything but an enamel, glass, or china container. Despite her interest in and respect for things made by and distinctive to the groups in the area, hosts were more concerned with gracing the table with items that adequately reflected their view of status, as well as their own economic and social standing.[3] The same gesture is made in indigenous social situations where offering food is a crucial part of hospitality and congeniality. Even in the Ga'anda area where most people have rela-

144. A display of enamelware containers in a Bole woman's room (*bing*); Ribadu. July 1982.

145. Interior view of a Tera woman's room (*kɇba*) showing stacked pottery and decorated gourds; Shinga. See Plate 9.

tively little in the way of material wealth, women serve food in enamel bowls whenever their husbands invite guests for a meal. When a man is alone, however, he still prefers to eat from traditionally decorated calabashes. Indeed, most men and women say they prefer the taste of food and liquids taken from gourd containers and consider the way metal bowls "sweat" to be disagreeable. In addition, many people find decorated gourds more aesthetically pleasing and thus, still able to enhance the experience of eating or drinking. The continuing use of decorated gourds seems, subtly, to reflect pride in local cultural accomplishments and a determination to preserve them, despite incentives to modernize.[4] This double standard helps explain why gourd cultivation and decoration have remained viable in rural communities regardless of the influx of mass-produced containers.

The material syncretism of modern Nigerian life may be due to the prevailing impulse—whether people live in rural or urban communities—to recognize the intrinsic advantages of things and materials. On a practical level, newness or oldness is unimportant—there is a place for advanced products of Western technology, and there is still room for items made by hand from available raw materials (Fig. 147). Trappings of modern life have a usefulness even in remote villages where they seem incongruous, and certain traditional household objects are not yet obsolete in towns. By combining elements of both worlds, people can embrace what is new while still acknowledging and benefiting from their own distinctive traditions.

A woman in Ribadu proclaims her elevated social status with an array of enamel containers placed next to a traditional pot-stand surmounted by a large decorated gourd bowl (Fig. 144). The fact that the latter serves as a symbolic reference to past tradition does not dilute the message of the modern display. Nor is there a contradiction in using the clay vessel as a support for a large cardboard box, an enamel tray, and a bolt of cloth wrapped in plastic. The tendency to make maximum use of any item or material, once so true of gourds, is now particularly evident in the way people save, reuse, and substitute modern "throwaways"—empty tin cans, plastic wrappings, styrofoam molds, plastic kegs, cardboard boxes and tubes, and metal drums. Although it is encouraging that people have remained so resourceful, it is also true that it is much easier to pick up, for example, a discarded plastic jug and use it as a water container than to take the time and the effort to grow, dry, clean, and possibly decorate a calabash bottle. Nonetheless, this ingenious exploitation of imported materials and objects does not preclude the continued use of traditional ones.

Gourd decoration has also been adapted to the spirit of modern life. New images have been adopted to reflect social, political, and religious change (see Pl. 32). The representation of things like helicopters, airplanes, and trains not only demonstrates a familiarity with these emblems of modernity, but by extension makes the gourd's artist and owner participants in the larger world they symbolize (Fig. 148). The same is true of prayer boards—visual symbols of the status and power associated with Islam in northern Nigeria (Fig. 149). The fact that gourds were traditionally decorated with motifs inspired by the environment and local material culture means that today, the designs they carry reflect the pluralism that has replaced the individualism once fiercely guarded by most peoples of northeastern Nigeria. The motifs engraved on a large gourd bowl photographed in a Waja village reveal an unusual patchwork of cultural associations—the bands on the rim are Waja, the complex pictorial array of animals, birds, insects, prayer boards, and human figures is Fulani, and the Roman script is Euro-American (Fig. 150).[5] As a unique composite statement, it demonstrates the potential for individual, as well as group, cultural expression.

The inherent adaptability of gourds to many modes of decoration has allowed them to keep in step with technological changes. Alhaji Bundi of Gombe is credited with introducing to the Gongola Valley the technique of decorating the outsides of gourd bowls and spoons with imported enamel paints.[6] Over the last fifteen years he has taught a number of men this relatively simple procedure and each has carved out a particular territory where he goes, like a traveling salesman (all his tools and materials in one wooden case), to weekly markets in different towns and villages. Berns watched Mohammadu Gombe decorate gourds at the Friday market in Dumne, the main town in the Yungur district (Pl. 38). All day long, women brought him plain gourds that he painted in a quick and efficient assembly-line fashion: the tops were covered with one vibrant color; after they dried, the rim edges (of bowls) were painted with a second color; and then the tops were printed with contrasting patterns using stamps made from the multilayered soles of rubber sandals (zoris). Designs were carved or burned into the rubber so that on contact, a configuration of small dots was produced.

146. Gifts of sorghum, benne seed, and cloth presented to the family of a deceased Yungur elder during a Wora second funeral ceremony; Dirma. May 1981.

147. Man riding a bicycle, wearing a wristwatch, and carrying a gourd bowl; Potiskum. 1970.

148. Pastoral Fulani pressure-engraved gourd with airplane motif. Purchased in Rakuba, near Jos. 42 cm x 46.5 cm. Herbert M. Cole collection.

149. Pastoral Fulani (Jafun?) pressure-engraved gourd with prayer boards. 30.5 cm x 36.7 cm. UCLA MCH X83–803.

150. Waja woman pyro-engraving a gourd bowl and displaying a prized example from her collection; Dela Waja. February 1982.

The simplicity of this technique makes it possible for Mohammadu to paint as many as two hundred gourds a day. Predictably, each gourd looks very much the same, with only minor variations in the stamped designs. Yet the popularity of this mode of decoration among the Yungur is curious (cf. Pl.12; Fig. 146), given the technical precision and stylistic detail characteristic of their pressure-engraving and pyro-engraving (cf. Figs. 52,84). Women do not seem concerned that unlike the relative permanence of most traditional decoration, the inexpensive enamel paint is thinly applied and wears away easily (see Pl. 12). They prefer it because it is a very fast and cheap form of decoration—from ten *kobo* ($.15 U.S.) for a spoon to thirty *kobo* for a large carrier-container. Even more appealing to the Yungur and to the other groups for whom this service is available, may be the visual parallel between painted gourd bowls and enamelware basins and containers. They have the same glossy, brightly colored surfaces and repetitive designs. Even the colors used on gourds—red, blue (or black), and white—predominate on enamel bowls and basins. Thus, traditional calabash bowls are given some of the aesthetic appeal, as well as the expressive meaning, of mass-produced prestige containers.

That this form of decoration was motivated by entrepreneurial gain is unquestionable. No doubt Alhaji Bundi was responding either to the success of male Hausa traders who sell their goods all over northern Nigeria (and beyond) or to merchants with open stalls in the markets of larger towns and cities like Gombe (Pl. 4). The same economic expediency may explain the way local female artists have reduced traditional compositional formats to summary statements so a greater number of gourds can be sold more cheaply at markets (Fig. 74). These less elaborately pyro-engraved gourds are usually sold in town centers characterized by a mixed ethnic population, such as Bauchi, Gombe, Biu, Shani, Potiskum, and Yola.

Decorated gourds usually sold by men in open markets are also popular among Westerners who value them as examples of "traditional" craft. The tourist and expatriate market in Nigeria has provided an additional commercial incentive for the continuation and expansion of this art, especially for groups like the Hausa and the Kanuri, and the Yoruba in the south. Additionally, a concentration on recognizable representational imagery shows the artists' sensitivity to the aesthetic preferences of their Western customers (Fig. 151; see Pl. 33). Even though these gourds may differ from the

151. Hausa gourds (*k'warya*) purchased in northern markets. a. Gourd bowl. 21.6 cm x 22.9 cm. UCLA MCH X85–58; b. Gourd bowl with airplane motif. Russell and Maxine Schuh collection; c. Gourd bowl. 18.1 cm x 19.5 cm. UCLA MCH X85–57; d. Gourd wall plaque. 22.5 cm. Berns collection; e. Gourd cup (*k'ok'o*); f. Gourd spoon (*ludayi*) with scorpion. Russell and Maxine Schuh collection.

work of village women that is generally unavailable for popular consumption, they reflect the dynamism and resilience of the art, as well as the resourcefulness and adaptability of the Nigerian entrepreneur. In fact, the alteration of gourds to suit a foreign market can be even more drastic—pieces of gourd shell have been cut into a variety of shapes, decorated, and then sold as wall plaques (Fig. 151); hemispherical bowls have been carved with fancy openwork designs and thus become suitable for decorative lampshades.[7] Indeed, it is not unusual to find decorated gourds of all shapes and sizes in the houses of expatriate residents and Westernized Nigerians, purely in the form of ornaments.

Modernization in Nigeria has provided many replacements for gourds. It also has inspired new images, uses, and markets for this ancient artistic tradition. That all the groups represented in this collection were still producing and decorating gourds in the early 1980s is further proof of their enduring popularity. Gourds remain a cheap, useful, convenient, durable, and in many instances, preferred container. The traditional art of decorating them is still the most vigorous in rural areas where they not only serve a domestic purpose, but also have meaning in social, ceremonial, and symbolic terms. Perhaps this monograph will stimulate and inspire the people whose gourds now grace the walls of our museums and our homes to continue exploiting this versatile medium, if not because of its inherent utility, then because of its potential as a vehicle for preserving local identity and for demonstrating technical achievement and creative verve.[8]

APPENDIX
Word Lists by Language Family

I. WEST ATLANTIC

A. Fulani[1]

balere	*kechere* that has been stained black
bodere	*kechere* that has been dyed red after the designs are engraved
fefande	decorated gourd (pyro-engraved)
futere	gourd with a "warty" exterior, sometimes decorated on the inside and with the textured exterior constituting a natural decoration
gimbiyare	small drinking gourd with inner seal of black or red pigment, sometimes also decorated on the outside
gumbal	whole gourd with a hand-hole cut in the top, used as a water-carrier
hasere	*kechere* with interior painted designs for drinking liquids
horde	gourd spoon or ladle
jalbal	pressure-engraving needle
janturu/junguru	tubular gourd, cut open at one end, used for applying henna or pouring sauce
jolloru	bottle-gourd with the top sliced off for drinking water
kechere	medium or large-sized hemispherical bowl with pyro-engraved designs
lairu	small gourd pot with a hand-hole for storing pomades and oils
tummode/tummude	gourd (generic)
tummode daneji	undecorated gourd (white)
tuppande	decorated gourd (pressure-engraved)

II. CHADIC

II-A. WEST CHADIC

A. Hausa[2]

butulu	narrow portion of the *shantu* used as a blowing horn
buta	bottle-shaped gourd
duma	gourd (generic)
gako	gourd spoon with solid neck
gora	very large spherical gourd used as a float to cross rivers
gyandama	bottle-shaped gourd
jallo	pear-shaped gourd to carry water for ablutions
jemo	a short wide-mouthed, club-shaped or pear-shaped gourd used to hold milk or liquids
k'ok'o	small calabash bowl
kololo	gourd spoon or ladle with hollow neck
kumbu	small gourd bowl with a cover
kurtu	small globular gourd for holding milk or seeds
kurtun tawada	ink pot (usually tubercled)
k'urzunu	tubercled ("warty") gourd
kwaciya	small "cup-sized" gourd
ludayi	gourd spoon or ladle with narrow neck
k'warya	any circular gourd
mabakaci	largest spherical gourd also used to separate grain from husk
masaki	largest spherical gourd used as a dye vat or as a float
mod'a	larger bottle gourd with curved neck that serves as a handle, the body pierced for use as a dipper or handled pot
shantu	long gourd trumpet (aerophone)
wuk'a	decorating tool (knife)
zunguru	cylindrical gourd for applying henna to the hands

zuru	cylindrical gourd for applying henna to the hands

B. Ngizim[3]

ad'iyu	gourd plant (generic)
ciizemi	long gourd, used to store small things
dlagwdaru	old broken bowl
fena	hemispherical eating bowl
gadakuwa	small bowl for eating
geruwa	gourd bottle
gwabo	large spherical gourd
kambi	small bowl for storage
kurtu	small round gourd
mabu	very large bowl
zunguru	long gourd, for application of henna to hands

C. Dera[4]

lib'e	gourd (generic)
lib'e buni	gourd for flour
lib'e b'uta	gourd for water
lib'e d'ila	gourd for soup
lib'e d'oyara	gourd for porridge
lib'e kayo	decorated gourd
lib'e muni	gourd for porridge
lib'e tib'e kokumat	gourd cover
war kayo	decorating tool (pyro-engraving)

II-B. BIU-MANDARA

A. Ga'anda[5]

buta	bottle-shaped gourd
kapatib'a	largest gourd measure
ketere	gourd pulp
kowata	small bowl for eating soup
kwerte	gourd for measuring out grain
mafatib'a	gourd measure, large size
sambata	undecorated gourd drinking bowl
teb'fenda	gourd to measure out grain for grinding
teb'fuxwada	gourd bowl for grain
teb'kennda	gourd for giving husband beer in the morning
teb'sayema	bowl for drinking water
teb'tarta	gourd to dip out corn from granary
teb'waa	small bowl for children
teb'wanketa	gourd to measure out grain equivalent to one hoe blade
teb'yata	bowl for eating porridge
tib'a	gourd (generic)
wanb'eleta	gourd spoon
wankowata	small ornamental bowl
wantib'e	gourd on the vine
xafta	decorating tool ("arrow point")
xed'ititibe	undecorated gourd

B. Hona[6]

d'engcerehla	decorated gourd
d'engkyaxena	gourd for soup
d'engnda	gourd (generic)
d'engse'ama	gourd for water
d'engsub'yara	gourd to cover porridge
d'eng'yara	gourd for porridge
kubara	large gourd for water
shara	gourd spoon
shar'teha	gourd spoon for porridge
xafira	decorating tool (pressure-engraving)

C. Tera

belangb'eri	large grain gourd used by a new bride
deb'a	gourd (generic)
dekun	gourd for porridge
divina	gourd for keeping flour and hanging from rafters
diviyin	gourd sun/rain hat
kehla	bottle gourd
shereh	gourd spoon
sud'ida	decorating tool
sunkur	gourd henna tube
yeka	small decorative gourd
yekazag'yem	water gourd

D. Pidlimndi

d'erba	gourd (generic)
d'erba d'ektigunu	gourd for soup
d'erba d'ektijem	gourd for water
d'erba turogundiyeti	gourd for porridge

III. ADAMAWA

A. Longuda

kelawa	gourd (generic; Dumna Zerbu dialect)
kwarawa	gourd (generic)

B. Waja

bela kaage	decorated gourd
bokotunge	small gourd for drinking water or eating soup
kaage	gourd (generic)
kaagopopulo	undecorated gourd
kaawaips	gourd carry-container
kamwone	gourd for porridge
soro'ali	gourd for soup
tummeh	gourd for flour
yau	decorating tool (pyro-engraving)

C. Gbinna[7]

b'aa yor kõjeba	decorating tool (pressure-engraving)
b'aa kõsa	decorating tool
domlomra	gourd spoon (Tenna)
kõgeno	gourd for soup (Tenna)
kõkete	gourd for porridge (Tenna)
kõtoto	gourd for porridge (Riji)
koxõketawa	gourd for porridge (Yang)
kõ yorsa	decorated gourd (Riji)
kwake	gourd for soup (Riji)
tagene	gourd spoon (Yang)
yoroma boma	decorated gourd (unblackened)
yoroma teptep	gourd for soup (Yang)

D. Yungur

bwa kũsa	decorating tool (pyro-engraving)
dinge	generic (gourd)
dinge dore	decorated gourd
kũsa	gourd (Suktu dialect)

IV. JARAWAN BANTU

A. Mbula

gilu kwar	decorated gourd
kwar	generic (gourd)

NOTES

·1· INTRODUCTION

1. The words "gourd" and "calabash" are used interchangeably in this text. Both terms refer to the fruit of the flowering plant, *Lagenaria siceraria*, considered native to Africa. Botanically distinct, but identically named, is the fruit of the tropical tree, *Crescentia cujete*, indigenous to the New World. This confusion of terms is evident in the literature (Chappel 1977 Heiser 1979, Jeffrey 1967, S. Price 1982) and is reflected in common usage.

2. References to the early descriptions of gourds by travelers were drawn from Sieber's survey of calabashes as household equipment (1980:178,195). He, too, includes this passage from Jobson (1968:168–169).

3. See Astley (1745–1747 I:331) and Burton (1860:314–15).

4. One remarkable but not widely-known organization, The Gourd Society of America, now reformed as The American Gourd Society, has, since the 1940s, disseminated information about gourd growing and gourd craft in the United States. A recent book on gourd craft suggests persistent interest (Mordecai 1978).

5. See the annotated bibliography compiled by L. Pharr, "The Ubiquitous Gourd: The Art of Calabash Carving in Africa" (1983).

6. E.g., Fagg (1972); Gardi (1969); Hodge (1982); Jefferson (1973); Newman (1974); and Price (1975).

7. E.g., Gabus (1967); Griaule and Dieterlen (1935); and Herskovits (1938).

8. See Chappel (1977).

9. E.g., Chappel (1977); Hodge (1982); Konan (1974); and B. Rubin (1970).

10. While such "shows" may be motivated by political interests and the exploitation of "local color" by a ruling Fulani elite, they nevertheless have helped to reinforce local pride in traditional craftsmanship.

11. Perani (1977:9) makes this observation in her review of Chappel (1977), noting the importance of gourd decoration in an "art impoverished" area (see note 19). While gourd decoration may be the most durable form of artistic expression, other arts are and were traditionally made by these groups—especially the Yungur and the Bata—such as figurative ceramics, body scarifications, brass ornaments, iron regalia, and decorative architecture. Although this conclusion on Perani's part may be a misreading of Chappel, who does illustrate other examples of art made in the area (1977:202–207), it does reveal the problematic distortions that can result from a study that focuses singularly on one mode of artistic production.

12. Detailed reports on physiography, geomorphology, climate, and other environmental features are contained in Aitchison et al. (1972). The major zones of relief described here are drawn from an introductory essay written by Bawden in Aitchison (1972:43–45) and from his Text Map 2. More general information on the geographic regions of Nigeria are provided in Udo (1970).

13. These figures are taken from Aitchison et al. (1972:text map 3).

14. This classification is discussed in Aitchison et al. (1972:16).

15. Davies provides some information on the evolution of political centralization in northeastern Nigeria, although many of his sources are questionable (1954–1956). A. Rubin considers the area around Biu to have been a primary "gateway" between Borno and the Upper Benue Valley (1974:161).

16. According to Chappel, gourd decoration appears to be confined to women of the Wodabe branch of the pastoral Fulani (1977:40). The Jafun and Daneji do not decorate their

own gourds, instead mostly commissioning them from the Wodabe. However, a few male Jafun and Daneji carvers were documented in the area around Gombe.

17. Greenberg (1966) provides the basis for the division of African languages into families; however, more recent scholarship has revised and superceded some of his internal classifications, including Newman (1977) and Hansford, Bendor-Samuel, and Stanford (1976a and 1976b).

18. More information on the influx of the Fulani, Hausa, and Kanuri into the northeast is provided in Chapter 5, see pp. 143–144.

19. The characterization of this part of Nigeria as "art poor" still persists among a number of scholars who base their definitions of "art" on a comparative ratings scale, influenced in part by a Western aesthetic standard valuing large-scale wood sculpture. For example, in his brief survey of the traditional arts made by the Margi, Vaughan indicates that the "Marghi live in an area of Africa which might be called art impoverished since they produce nothing comparable to the products of most of West Africa" (1975:184). While Chappel also subscribes to this general view, he has concluded nonetheless that the decoration of gourds, along with elaborate programs of body scarification, should be considered "major graphic systems," despite the fact that their producers and consumers may not regard them in this way (1977:176).

20. The earliest reports on the groups living in the more remote areas of northeastern Nigeria were filed by British colonial officers and are now deposited in the National Archives, Kaduna. Historical and anthropological information was gathered in order to develop the best strategies for implementing programs of indirect rule, especially in areas where such centralization had not existed before. The major published works that resulted from these observations are Temple (1919) and Meek (1925; 1931). For studies focusing on particular traditions, see Neaher (1979); Neher (1964); A. Rubin (1973; 1974); Sassoon (1964); and Wente-Lukas (1977a; 1977b).

21. Chappel worked in the Adamawa area from 1965–1966 (see 1973,1977). Arnold Rubin did brief reconnaissance work among a number of groups in 1970 as an extension of his comprehensive survey of the arts of the Benue River Valley. The results of Rubin's research prompted the more intensive and extensive investigation of these art-making activities by Berns in 1980–1982. Rubin's work in the Lower Gongola is reported in his *Sculpture of the Benue River Valley*, a forthcoming publication of the UCLA Museum of Cultural History. We are grateful to Dr. Rubin for making a copy of his manuscript, as well as his field notes and photographs, available to us for this project.

22. Leith Ross' *Nigerian Pottery* (1970) provides some indication of the degree of variation and skill evident in the production of utilitarian, and to a lesser degree, ritual pottery in northeastern Nigeria. Other brief surveys and articles have been published on local or regional ceramic traditions, see Chappel (1973); Gauthier and Jansen (1973); Kandert (1974); Krieger (1965–69); Paulme-Schaeffner (1949); Pearlstone (1973); Slye (1969, 1977); Stevens (1976); Teilhet (1977–1978); Wittmer and Arnett (1978).

23. This reference was taken from Sieber (1980:18), who in turn drew from the excavation reports by M.D. and L.S.B. Leakey (1950:pls. II,IX).

24. See Shaw (1977) for a highly readable narrative account of the Igbo Ukwu excavations, which earlier were reported in an exhaustive two-volume compendium (1970). For a brief overview of the shorter account and the issues it develops, see Berns (1978a). Also see Cole and Aniakor (1984:18–23) for a good summary of what is known about Igbo Ukwu and how it relates to the arts and culture of modern Igboland.

25. For a clear explanation of lost-wax casting techniques, see Shaw (1977:15–17).

26. Lebeuf considers all cuprous materials excavated in the Sao area to be objects of regalia, found exclusively in burial contexts (1982:31–35). For more information on ancient metallurgy in the Lake Chad Basin, see Lebeuf (1982); Lebeuf and Françaix (1978); and Hamelin (1952–1953).

27. Other West African archaeologists have criticized the Lebeufs' periodizations and regional sequences based on more recent excavations and dates from Nigeria and Chad. See David (1976:234); Connah (1976:321); and Shaw (1969:193). For a summary of the Lebeufs' most recent report in French on the Sao corpus (1977), along with some brief comments on its critical reception, see Berns (1978b).

28. Chappel (1977:36) has based this number on the words listed in the Fulani dictionary compiled by Taylor (1932) and with special reference to the gourd types used by the settled Fulani in the Adamawa area. The Appendix provides lists of words used by the Fulani and many of the other groups represented here, mostly field-collected by Berns.

29. We are grateful to Dr. Russell Schuh, Department of Linguistics, UCLA, for his insights on the linguistic implications of the noncognatic words used for gourds.

30. These two words are found in English, French, and German. In Spanish, only the word *calabaza* (calabash) is used to refer to the fruit of the New World tree (*Crescentia cujete*). For comments on the confusion in nomenclature and species, see Price (1982).

31. Mercier's definition of "calabash" in the *Dictionary of Black African Civilization* includes a reference to Dahomean (Benin) cosmological thought strikingly similar to that of the Hausa: "The universe is a sphere consisting of two halves of a calabash—one half, holding the sky, is turned upside down and rests on the other, containing the waters and the earth, while the circle where the two halves are in contact is the horizon" (1974:76).

32. The first seven proverbs listed here are cited directly from C.E.J. Whitting who includes the Hausa, its literal translation, and an explanation (1940:10–14,117). The remainder have been extracted from R.C. Abraham's (1962:593–594) lengthy citation of the literal and idiomatic usage of the word *k'warya* ("any circular gourd" bowl); verbatim translations of his Hausa proverbs were done by Berns. Some of Whitting's explanations also have been reinterpreted in light of Abraham and in consultation with Russell Schuh.

33. Skinner (1968:86) includes this same proverb, but considers it equivalent to our "Let bygones be bygones."

·2·

AN ETHNOGRAPHY OF GOURD USE

1. The Argungu festival (near Sokoto) is the most well-known. Konan indicates that although gourd fishing floats are never decorated, the event provides artists with an opportunity to sell gourd bowls to visitors, decorated with motifs depicting the events of the day (1974:8; fig. 15).

It has also been noted that "ferry boats" are built with gourds (Arriens 1928:149). Globular gourds suspended under a wood lattice-work raft accommodate passengers and their cargo.

2. Despite the unusual shapes of bottle gourds, it is unlikely that cultivators interfered with the growth pattern of gourds. There is evidence, however, that a bottle gourd weighted while on the vine will grow a more attenuated neck.

3. See Chapter 4 on the Ga'anda for more specific details on how the gourd is bisected and cleaned.

4. Chappel has suggested that bottle gourds were depithed in this way (1977:8). Sieber (1980:178) cites Astley's early observation of how the inner pulp of a bottle gourd was removed:

> To prepare them for Use, they made a Hole of a proper Size near the End, into which they pour warm Water, in order the sooner to melt and dissolve the contained Pulp (Astley 1745–1747 II:331).

5. During his work in the Adamawa area in the mid-1960s, Chappel observed that it was mainly in transactions centering on the proposed purchase of gourds for museum collections that monetary compensation was a major issue (1977:24).

6. Interview conducted with Dambaya in the town of Fika, January 1971, by Hudson.

7. Interview conducted with Bulama in the Margi town of Balala, January 1971, by Hudson.

8. See Appendix for word lists by group.

9. The wide application of the word "*k'warya*" is given in Abraham (1962:593–594).

10. An example of this type of gourd from Zaire is illustrated by Sieber (1980:198).

11. Among the Banyankole (Nkole) pastoralists of Uganda, women put the little amount of milk that can be spared each day into a bottle gourd. The mouth is then corked and the gourd rocked back and forth either on a grass pad or on the knee until the butter separates (Roscoe 1915:108). The same group also uses a variety of straight-necked bottle gourds, which achieve a lustrous dark patina, as milk pots. Roscoe writes that only gourds, wooden bowls, or clay pots are suitable for holding the milk provided by their herds (1915:106). The Nkole believed that other containers would be injurious to the cattle, an attitude similar to that held by the Fulani (see below, Chapter 3, note 31).

12. See Northern (1975:fig. 33). Elaborate beaded versions will be discussed briefly below, p. 94.

13. Masai cattle herders in Tanzania also use long cylindrical gourds, often ornamented with leather and cowries, to catch the blood that is drawn from a cow's neck. An arrow pierces the jugular vein and the blood, mixed with milk, is given to newly circumcised children (Saitoti 1982:88–89). See also Seiber (1980:206–207).

14. Dunhill (1969:11–130) discusses the distribution of water pipes and includes a number of sketches of pipes from Tanzania, Zambia, Zaire, and Angola that show the various ways the gourd has been adapted to this purpose and how its ornamental potential has been exploited (see figs. 152–169). The custom of cooling and cleansing the smoke of a pipe by drawing it through a vessel of water appears to have been developed for smoking both hemp and tobacco. In fact, most people associate the word "calabash" with a popular type of Western pipe that is fashioned out of the curved neck of a gourd. The gourd is ideal for this purpose because of its light weight, its large air space that yields a cooler smoke, and its tendency to develop a rich patina (Weber 1962:36). Weber claims that the use of gourds for making pipes was discovered by the Dutch when they founded Cape Town in 1652 (1962:35). While the African cultivator rarely molds or trains the gourd into a specified form, relying instead on the many varieties that have evolved through natural selection, the curve of the neck on a calabash pipe made in the Western world has been individually controlled:

> While a gourd is growing, the cultivator aids in the formation of its gracefully curved neck by gradually training the neck to the correct form. This is done by placing under the gourd a flat board in which pegs are inserted, pegs that hold the neck in a prescribed position. These pegs are then moved, little by little, so as to force the neck into the curve desired (Weber 1962:36).

15. Chappel explains that gourds were traditionally produced by Yungur women who used the technique of pressure-engraving (1977:46). It is only in the Diterra area that men have taken over this once exclusively female occupation. It is worth noting that, of all the Yungur sections, Diterra is the most accessible by road and is dominated by the large town of Dumne where the Sudan United Mission (now the Lutheran Church of Christ, Nigeria) was established in 1933. This fluid context of changing ideas may have made it easier for such a role reversal to take place. The technical dichotomy in Yungur gourd decoration is also unique. In Diterra, men do pyro-engraving while women still do pressure-engraving. However, in Suktu and Waltandi (the two other major Yungur sections), both techniques are only practiced by women.

16. Most of this discussion of Yungur marriage proceedings was drawn from Chappel (1977:14). His account and most of Berns' supplementary material focuses, for the most part, on the Yungur (who Chappel calls the Pirra Yungur) who live around the town of Dirma. According to Berns' later information, the name preferred by the Yungur for this primary occupation area is Diterra, corresponding to the sacred hill

site where the main chief (Gubo) traditionally lived and maintained his shrines.

17. The following description of Fulani practices was drawn from Chappel's (1977:10) account; he in turn relied on Stenning (1952) and Dupire (1963).

18. Although this elaborate mode of household display is distinctive to the Tera, it should be noted that the convention may have been adopted from their Bole neighbors who maintain a similar, but less complex, tradition emphasizing the display of decorated pottery. The historical implications of the parallels between these two groups will be discussed briefly below, Chapter 5, pp. 151–152.

19. See also B. Rubin (1970:fig. 20).

20. For more on "Adamawa Calabashes and a Pastoral Fulani Aesthetic," see P. Mauk's unpublished master's thesis (1980).

21. See Chappel for drawings of Fulani facial scarifications (1977:fig. 219a-e).

22. In a parallel context, the neighboring Ga'anda use a gourd armature to make face masks (*magenshen*) worn during Sapta, the boy's initiation ceremony. Features are either cut out of or built up onto a calabash bowl that is covered with beeswax and studded with lines of abrus seeds; ram's hair is used for whiskers and eyebrows, and cow's teeth are inserted into the open mouth. The masks are worn to frighten initiates who are told that "spirits" are dancing.

23. This Dadiya dance rattle was documented and collected by A. Rubin in the village of Dadiya Dutse in 1970. Berns recorded two identical rattles, along with one detached handle in a neighboring Kwa village on the riverain side of the sandstone hills that flank the Benue (the Dadiya live on the opposite flank of the same chain of hills). Called *kishibiya*, they are kept by Kwa war priests and used in celebrations honoring men who demonstrated great bravery in war or more recently, skill in the hunt.

24. This illustrated iron dagger was collected by A. Rubin in 1970 in the village of Tula Wange, just northeast of Dadiya. It is identical to the *nyansanye* documented by Berns in the village of Dadiya Dutse in 1982. The correspondences in certain categories of ritual objects—such as iron regalia—among the groups living in this Benue-Gongola confluence area are supported by their close linguistic affiliation.

25. This Fali funeral ceremony is illustrated in Huet (1954: fig. 29) and Huet et al. (1978:figs. 177–79). An even more dramatic two-page illustration of this same festival, where a mass of women and girls wind around a single male dressed in Fulani robes, is reproduced in *The Horizon History of Africa* (Boahen et al. 1971:302–303). While the women dance, the male spectators provide the rhythm by constant clapping. Huet's impressive photodocuments of African dance (1954; and Huet et al. 1978) include other examples of gourds used as dance attributes as well as musical instruments.

26. It should be noted that, among most groups represented here, women were traditionally restricted from cooking the game their husbands killed during a hunt. Many men still own tripod-based cooking vessels in which they prepare certain types of meats only they consume.

27. If no deaths have occurred, Kwefa must be held independently by each Ga'anda community at least once every seven years as a way of ensuring health and prosperity. It also must be performed in the year the boys' initiation (Sapta) is held, which also follows a seven-year cycle.

28. Neither Meek (1931 II:464) nor Chappel (1877:20) offers an explanation for why a plain gourd operates as a ritual substitute for an ancestral skull. In fact, Meek describes a Yungur healing procedure that directly involves the use of an ancestral skull (1931 II:464). A tooth is removed from the exhumed and enshrined skull, ground into a powder, and then administered, mixed with beer, to the patient to effect a cure.

29. See Nissen (1968:130) for more information on Ngala. An unpublished manuscript prepared by a local Mbula education officer, Mr. Yustus Offah (n.d.), explains this society and other aspects of Mbula traditional life in considerable detail.

30. The stepped cement platform built over this grave appears to be a local innovation. It substitutes for the traditional Mbula grave dug under the floor of the deceased's sleeping room. In the case of the latter, the round mud room is elaborately decorated in relief designs and then painted and ornamented with screens of woven mats and fringed raffia. The temporary shed erected over the cement grave is instead constructed in front of the room, similarly protecting the display of ritual materials.

31. Unfortunately, what happens to most of the personal belongings kept on a grave was not recorded. The two large stones bound with fiber kept at the base of the shrine each represent a large animal (a leopard or a bush cow) killed in the hunt. These stones are kept together in each hamlet to serve as permanent reminders of male bravery and skill. Red ochre is poured over them during preharvest festivals to activate the powerful spirits they represent.

32. This is only one of a number of techniques used by the Yungur *sife* who is considered able to cure not only physical illnesses but psychological disorders caused by the interference of malevolent spirit forces. Nissen provides a short section on the important role of diviners among the Yungur (1968:149–150).

33. These "spike" figurines represent a rare category of wood sculpture found north of the Upper Benue River. The pair illustrated here by the Burak are very rudimentary carvings that appear to be copies of those made by their Pero neighbors. Unfortunately, no appropriate illustrations of the Pero versions were available. Related pairs of wood figurines have been erroneously attributed to the Waja, among whom no carved figures of this sort have been documented. Recent research by A. Rubin indicates that they were made by the Wurkun, who speak a Pero dialect and who use them as "field guardian" figures (1976:25,fig. 17; see also Sieber and Vevers 1974:24). Information Berns collected among the Pero and Burak, who live on the opposite side of the same sandstone hills as the Wurkun, suggests that these sculptures also may have been used by the Wurkun in the divination context described above (see also Coppens n.d.). The spike attachment

helps prevent termite infestation when kept enshrined in the diviner's room. Pero carvings resemble examples of Wurkun "pole-figures."

34. Information about both these divination procedures was collected in the field by A. Rubin (pc:December 1984).

35. An interesting example of this expressive role of gourds is found among the Yoruba who use calabashes of fortune (*igba iwa*) during the installation of a new king (*oba*) in order to predict the future events of his reign (Ojo 1966:246–247). Two gourds with lids of the same shape and elaborate decoration were presented to the *oba* who had to choose one. If he chose the one containing jewels and other prestige materials, it indicated that his reign would be peaceful and prosperous; if he chose the other, which held swords and arrows, it was symbolic of war and unrest.

36. Olive Macleod, in *Chiefs and Cities of Central Africa* (1912:91), describes a similarly configured long gourd flute from the area of the Tuburi Lakes among the Wadama people. She reported that the man who played it "was of great blood, and he alone of all his race was able to play" it. Like the Gbinna example, it also was made of four interlocking parts. Her description of its construction warrants repetition here:

> The sections are fastened together by means of cane pegs, and the joints are rendered air-tight by an external plaster of mud and cow dung, or a layer of some mucilaginous compound (decayed rubber?): in the latter case the joint is further concealed by a covering of hide.

37. See A. Rubin (1985:98–99; fig. 86).

38. Other related groups living in the Gongola-Hawal confluence area play similar fixed-key xylophones. Among the Bura and the Tera, they serve as the central components of secular dance bands. Almost everywhere, such xylophones are now constructed with cow horn rather than gourd resonators, which produce different sounds. Although only speculation, it seems likely that the incorporation of horns would postdate colonial pacification when they would have become available in sufficient numbers for this purpose. For more details about musical instruments used by northeastern groups—Ga'anda, Tera, Bura, Bole, and Tangale—see Newman and Davidson (1971).

39. Among the Bantu-speaking peoples of South Africa and Mozambique, gourd xylophones are considered to be one of the most outstanding examples of craft technology. They are highly developed among the Chopi who play them as large orchestras. Bottle-gourd resonating chambers and wood keys of varying sizes yield tonal patterns that can be differentiated as soprano, alto, mezzo-soprano, bass, and contrabass; each is given a different vernacular name. H.-P. Junod discusses the elaborate xylophones made by the Chopi of Mozambique as well as the Venda of the Transvaal area (1936:47–48; pls. LXX-LXXI,XV). He includes in his discussion a passage written in 1609 by João dos Santos who found a xylophone at "Quiteve's court." The description shows that examples found as far away as Nigeria, and 350 years later, are identically conceived:

> The best and most musical of their instruments is called the *ambira*, which greatly resembles our organs; it is composed of long gourds, some very wide and some very narrow, held close together and arranged in order. The narrowest, which form the treble, are placed on the left, contrary to that of our organs, and after the treble come the other gourds with their different sounds of contralto, tenor and bass, being eighteen gourds in all. Each gourd has a small opening at the side near the end, and at the bottom a small hole the size of a shilling [patacão—a coin of the time], covered with a certain kind of spider's web, very fine, closely woven, and strong, which does not break. Upon all the mouths of these gourds, which are of the same size and placed in a row, keys of thin wood are suspended by cords, so that each key is held in the air above the hollow of its gourd, not reaching the edges of the mouth. The instruments being thus made, the Kaffirs play upon the keys with sticks, after the fashion of drumsticks, at the points of which are buttons made of sinew rolled into a light ball of the size of a nut, so that striking the notes with these two sticks the blows resound in the mouths of the gourds, producing a sweet and rhythmical harmony, which can be heard as far as the sound of a good harpsichord.

Other information on the Chopi is available in P.R. Kirby (1934:57–65).

40. Information about the Chamba xylophone was provided by A. Rubin (pc:December 1984).

41. Another group known for their exploitation of gourds as musical instruments is the Manding. One of their principal instruments is a xylophone that has up to nineteen keys, called a *bala*. While it is played today by popular Senegalese ensembles, Manding oral traditions maintain that the *bala* owned by the great sorcerer/king Sumanguru was invested with magical powers and was played after each victory in battle (Dalby 1972:43). A twenty-one stringed Manding harp called a *kora*, is especially well-known. Said to have originated in Guinea, it is widely played by professional bards who use it as an accompaniment to traditional praise-singing (Dalby 1972: 43). A hemispherical gourd bowl, sometimes reaching an "immense" size forms the basis of the harp's complicated construction (Pevar 1978). Male bards, who are members of a hereditary class of musicians, play the primary instruments; Manding women sing and shake gourd rattles or perform on calabash drums (Dalby 1972:43). Rattles often are made by covering the gourd with a net that has been threaded with cowrie shells or seeds. The Manding also have an unusual "water drum," comprised of a small overturned calabash that floats on water inside a larger bowl. A brief inventory of Manding instruments is provided in Hambly (1935:429–431; pl. XCVII); also see *A Bibliography of the Arts of Africa*, compiled by D. Coulet Western (1975).

42. This summary has been drawn from two comprehensive surveys of Hausa musical instruments, Krieger (1968) and Ames and King (1971).

43. For a brief discussion of *shantu* music, see Mackay (1955).

44. Chappel came to a smiliar conclusion in his analysis of the social implications of gourd decoration (1977:22).

·3·

AN ETHNOGRAPHY OF GOURD DECORATION

Techniques, Designs and their Meaning

1. Chappel favors using the term "poker-work" for this technique (1977:34). However, the authors find this designation objectionable; it tends to conjure up the image of someone "stirring up a fire."

2. For a representative sampling of gourds decorated using additive techniques, see Sieber (1980:201–202,204–205,207–211). The most famous gourds in this category are those from the Cameroon Grasslands, where the gourd bottle is completely encased in a tightly fitting cover of elaborate beadwork. These will be discussed briefly below; also see Northern (1975; 1984).

3. See "Decorated Gourds and History," Chapter 5, pp. 143–162.

4. Although the gourd illustrated was made by a Bata-related group living in the Mubi area, Chappel used this description to characterize the heavy method of engraving developed by the Bata living around Yola (1977:44). Despite the geographical distance, the work of these Bata groups is clearly similar. In fact, Chappel contends that pyro-engraving was learned from the settled Fulani and that the Bata way of working is an innovative deviation (which has been taken up by the Mboi).

5. The composition of this Kanuri gourd is discussed further on p. 115.

6. Chappel identifies this same "rocking" technique with the work of the settled Fulani (1977:38).

7. This point is even better illustrated by a Kanuri calabash collected in 1967 by David Spain, reproduced in Sieber (1980: 200). It demonstrates a consistency in technique, as well as motif and composition, over the past fifteen years.

8. For examples of gourds worked in this technique, see Sieber (1980:203) and Trowell (1960:pl. LI).

9. These two Fulani styles of working are associated with differences in compositional format. They are described in Chappel (1977:44) and will be discussed below, p. 106 (see also note 28).

10. The Yoruba of southwestern Nigeria favor decorations carved deeply into gourds with thick walls. This technique enables the cuticle to stand out in striking relief, and often it is further stained with different colors to increase the sculptural effect. The deep carving can be executed with the knife held at an angle so that the beveled edges cast sharp shadows, adding further surface interest. For more on carved and scraped gourd techniques, see Trowell (1960:48); for illustrations of specific examples made by the Yoruba, see Trowell (1960:pl. XLVI), Sieber (1980:198–199), Price (1975:50), and Murray (1951).

11. In *Arts of the Hausa*, Heathcote (1976:46–47) includes a section on "Calabash Decoration" and explains that while carving may be the primary technique used, several other methods of working are common to Hausa men: pyro-engraving, dyeing, and applying white chalk. According to his catalogue (p. 47), some Hausa women also produce gourds with pyro-engraved designs. Perani (1985:6) also specifies that pyro-engraving is widely practiced by men in Kano state. In fact, one Hausa village forty km southeast of Kano, Kode, is a satellite craft center where the majority of men decorate gourds in this manner. They are bought locally or sold to traders who resell them in major northern markets.

12. According to Perani (1985:3), both the "whitened" and "reddened" gourds made by Hausa carvers are presented to Fulani women during marriage ceremonies. Indeed, it is Fulani patronage that provides the incentive for this Hausa decorative tradition and carvers often work in special sections of markets to better serve their select clientele.

13. This technique has been skillfully developed by various peoples living in Benin (Dahomey), particularly the Fon, who use their gourds to communicate through pictographic imagery. Trowell (1960:pls. XLVII-XLIX) illustrates a number of Bariba calabashes from Benin. The relief panels are covered with a range of imaginative zoomorphic images and some anthropomorphic motifs; bands of geometric motifs are carved in between. Based on Herskovits (1938), Trowell explains that the iconography represents symbolic stories, proverbs, or very often love messages; often these gourds are presented by young men to girls of their fancy. See also Griaule and Dieterlen (1935) and Herskovits (1938:344–353) for analyses of the symbolic content of their designs.

14. Information on Margi gourds was collected by Hudson in the village of Gulak in 1971. Though this way of dyeing may seem curious, Hudson's informants insisted this was the process they used. Vaughan (1975:185) indicates that oil and red ochre were used in this process, but red ochre would not seem to function as a dye.

15. For illustrations of gourds identified as *hasere*, see Chappel (1977:figs 153–165).

16. The Latin names for these plants have been taken from the references to the Fulani equivalents listed in Chappel (1977:38).

17. For more examples of resist-dyed *bodere* gourds, see Chappel (1977:figs. 133–152).

18. Although not represented in the Museum's collection, the settled Fulani also alter the exterior of an already pyro-engraved *kachere* by staining it black:

> The pods of the Egyptian Mimosa (*gabdi*) are boiled for several days in water. The concoction is then allowed to cool and the gourd is immersed for a few hours. Once removed, the surplus water is allowed to run off and the calabash, still wet, is buried in a large pot of black river clay or mud. It would appear that the concoction derived from the pods acts both as a primer and a fixative, for the black clay, after it has been scraped off, is found to have stained the surface shell of the gourd a deep black (Chappel 1977:38).

The interesting feature of this technique is that the stained areas of the unworked shell take on a smooth polish, while the engraved areas remain comparatively dull. Because the gourd (*balere*) is often totally black, the design emerges by way of textural contrasts. Sometimes, the impressed lines are filled with white chalk, which reverses the effect of pyro-engraving.

19. These drawings are housed in the Frobenius Institut, Frankfurt, and are part of the Ethnologisches Bilder Archiv; a number of these hand-drawn plates are published in Frobenius (1923). According to the Institut, Carl Arriens collected most of the gourds he painted in Kontcha, a village in northern Cameroon near the Nigerian border. As the watercolors of the gourds are dated 1911, this is probably the year of their collection. Arriens has written briefly about these gourds in a popular book on his experiences in West Africa (1928).

20. See David and Hennig for a brief description of the sedentary Fulani groups who established themselves in the Upper Benue region of northern Cameroon (1972:2–3). Their work was part of the Upper Benue Basin Archaeological Project and focused on the relationship between pottery and society. Frank Bartell, another member of the project, photographed decorated gourds made by the Fulani living in the area and sent examples to Hudson. They show that, like the Fulani living further west, pyro-engraving was the favored technique.

21. Wente-Lukas identifies a calabash now in the Berlin Museum für Völkerkunde [III C 29 365] as Verre (1977a:47; fig. 43). Berns studied this gourd in Berlin, said to have been collected during the Frobenius expedition in 1911. It is clearly of the same type as many drawn by Arriens. However, Chappel writes that the Verre have little use for decorated gourds, preferring instead pottery for domestic use and brass bowls for ceremonial purposes (1977:4).

22. In addition to their display value, these beaded gourds are also associated with the royal cult of ancestors. Northern explains that, among the Bamenda Tikar and the Bamum, they can be used as repositories for bone fragments of deceased Fon which are then displayed at the installation of a new Fon along with beaded ancestral figures (1975:119). The ornamental lavishness of palm wine containers parallels the social and sacrificial importance of palm wine.

23. E.g., the large Friday market in Gombi attracts women from Ga'anda, Hona, Gbinna, Fulani, Kilba, and Yungur communities. The weekly market in Biu provides a context in which Bura, Pabir, Tera, and Fulani women converge.

24. See Chappel's passage cited above for a description of how settled Fulani designs are constructed (1977:34). He also includes "roll-out" drawings of gourd designs that illustrate the range of pattern and motif used in settled Fulani compositions (figs. 3–44).

25. The Pidlimndi speak a dialect of Tera, which is classified within the Biu-Mandara branch of Chadic (Newman 1977: 5,38). Their Dera neighbors speak a language belonging to the West Chadic branch, suggesting divergent historical origins for the two groups. While this issue will be dealt with in more detail in Chapter 5, it is clear that the present-day Pidlimndi are culturally indistinguishable from their Dera neighbors. Thus, gourds collected among the Pidlimndi and the Dera are so similar that they can be discussed interchangeably.

26. For more on the relationship between the pastoral Fulani and Tera, Jera, and Waja gourds, see below, "Decorated Gourds and History," Chapter 5.

27. More detailed information on why gourd compositions used by the Tera and Bura are so closely related, as well as so various, will be provided in the next chapter.

28. Chappel associates the first mode of working with the Wodabe of the Ballala district, located south of the Benue near the town of Ribadu (1977:44; figs. 49–52). The second format is associated with the Wodabe of the Girei district, north of the Benue and the town of Yola (see Chappel 1977:figs. 54–60). The examples from the Museum's collection were purchased in Ga'anda (Fig. 80) and Peta.

29. Ga'anda compositions will be discussed in more detail in Chapter 4.

30. Chappel discusses at length the issue of image-making and the levels of awareness artists have about objects or ideas that have inspired their designs (1977:56–62). Using Arnheim (1967) as a model, he agrees that there is "little to be gained from a priority struggle as to what came first"—the form or the content (1977:56).

31. See Chappel for diagrams of Fulani facial scarifications (1977:fig. 219a–e). Unfortunately, only a limited number of referents are known for the complex motif systems on Fulani gourds. The difficulty in obtaining information from pastoral Fulani artists and clients about the meaning of gourd designs was stressed by B. Rubin (1970:25). She explained this reticence as an extension of the importance of cattle and their milk to the economic survival of the Fulani. As containers, gourds are thereby an "integral part of a complex involving charms and rituals to ensure the health and fecundity of the herd."

32. No Yungur gourds with this motif are included in this collection. See Chappel (1977:figs. 83–90).

33. Chappel notes that settled and pastoral Fulani carvers frequently use this conventional sign to depict a Moslem prayer board (1977:60; see fig. 195). He believes that such emblems can be "interpreted as a symbolic expression of the value placed on religious affiliation as a status determinant in the area." Likewise, visual signs interpreted as motor-cars may be symbolic of the value of these objects as they relate to male dominance and social and political status.

34. Perani indicates that such animal imagery has become a "hallmark of the Kode village style" (1985:6). While gourds decorated in this way are still used by the Hausa in traditional contexts, most of them are produced for the tourist market. These new directions in gourd craft, based on modern commercial incentives, will be discussed in Chapter 6 (see pp. 166–170).

35. Sieber illustrates an even more deftly executed example of a typical Kanuri equestrian scene depicting a caparisoned horse with its elaborately attired warrior mount and royal retainers (1980:200; see note 7).

36. The make of the car was determined by examining illustrations in Wyatt (1968) and Harvey (1985:24,52,82). The authors are greatful to Jack Carter for his information and expertise in this matter.

37. This information was collected by Hudson in the villages of Mazagani and Degubi.

38. Reconstruction of an interview held with the Chief of Askira, January 1971, conducted by Hudson.

39. Chappel has a chapter on "The appreciation of decorated gourds," that covers a range of perceptual contexts in which gourds are evaluated as artistic objects (1977:134–144). He also includes an appendix on "The aesthetic appreciation of decorated gourds" that details a ranking by observers of a standard collection of forty-five gourds (1977:144–172).

40. More information on these two artists and their work is included in Chapter 4 on Ga'anda gourd decoration.

41. Sieber in his discussion of decorated containers, recounts an experience he had in northern Ghana where a Lobi woman told him that she enjoyed drinking more if she used a decorated rather than a plain calabash cup (1980:167). It is useful to reiterate, however, a point made earlier by Chappel in this context, that the aesthetic value of gourds is also enhanced by the economic implications of their production, as well as their contexts of use.

42. A photograph taken by Vaughan in 1959 showing two Margi women seated together decorating calabashes illustrates his point (see Sieber 1980:197). However, it is often difficult to determine (if not so indicated) whether field photographs represent actual working conditions or the sort of demonstration staged for the benefit of the researcher.

·4·

THE GA'ANDA

Gourd Decoration from a Sociocultural Perspective

1. Field investigations were undertaken by Hudson for several weeks in 1970 and 1971 and by Berns from October 1980 through March 1981.

2. This settlement pattern makes accurate census counts difficult to obtain. Meek (1931 II:369) reported 5,400 Ga'anda; Kirk-Greene (1969:2) later recorded 7,641. Recent population density maps based on a 1963 census suggest these figures should probably be adjusted to approximately 12,000–15,000 persons (Aitchison et al. 1972:text map 12).

3. Meek (1931 II:379–386) includes a detailed account of Ga'anda social organization and marriage. Also see Hammandikko (1980:13–14) and Boyle (1916b:361–366).

4. The requirement that girls be scarified before marriage is noted in Hammandikko (1980:4), Meek (1931 II:384), and Nissen (1968:221). An interesting early article on this subject was based on Captain C.V. Boyle's firsthand observations early in this century (1916b). See Berns (1986b) for a lengthy discussion of Ga'anda scarification based on recent fieldwork. It should be noted that this practice was officially outlawed in 1978 by the Gombi Local Government authorities (Gongola State). There is little doubt that this proscription was an attempt to breakdown traditional social patterns and ethnic allegiances. The new brides who danced at the annual Xombata harvest festival in November 1980 (Pl. 34), had not completed the final and most elaborate stage of Hleeta.

5. Information about Sapta initiations included in Berns (1986a) is based on the oral accounts of Ga'anda elders in 1980–1981. The objective of Sapta is to teach boys three fundamental and interrelated skills: how to hunt and defend their future household; how to make tools and weapons associated with such activities; and how to endure the hardships they may suffer in discharging these responsibilities. During Sapta, initiates (*wankimshaa*) are flogged mercilessly with reed switches on at least four occasions. These whippings and a succession of other physical and psychological abuses test a youth's strength and endurance. Boyle (1916a) also witnessed this "Ordeal of Manhood." Brief descriptions of Sapta are included in Hammandikko (1980:12–13), Meek (1931 II:378–379), and Nissen (1968:219–220).

6. See Appendix for a list of Ga'anda gourd types. Also see note on Ga'anda orthography.

7. Information on gourd cultivation is taken from a short essay written for Berns by Musa Wawu na Hammandikko, who questioned local male cultivators in February 1981.

8. Information on this subject is based on explanations provided by Farin Jini (Mr. "Personality"), a male specialist living in Ga'anda Town. Berns also witnessed him preparing a harvested gourd in February 1981.

9. The training of artists was briefly discussed above (see pp. 94–95). Interviews conducted with five Ga'anda artists in Feburary 1981 by Berns indicated that learning began at about age seven, shortly after the first scarification markings were made. After three to four years of practice on broken gourds, a girl was deemed sufficiently skilled to engrave properly and to achieve satisfactory compositions.

10. For a description of Ga'anda pressure-engraving, see above pp. 80–83.

11. These names and those that follow are based on the responses of artists and patrons to questions about meaning asked by Berns and Hudson. Of all patterns named, those classified as filler motifs were most consistently identified in the same way. Other shapes were not always differentiated by name. Those named here correspond to informants' knowledge about specific examples.

12. For more on Kalhar, see below p. 133 and note 18.

13. This gourd was field-collected by Hudson in 1971, who recorded the name of the artist and the date of production.

14. The relationship between technique and design composition is discussed by Chappel (1977:184–185), who also uses diagrams to show how a typical composite design is built up.

15. The diagrams reproduced here are based on observations by Berns of three artists at work in February 1981. The same artists were also interviewed about their personal career his-

tories and working methods. There were clear consistencies among the five artists selected for more intensive observations. Yet, based on this limited sample, it is not possible to determine whether all Ga'anda artists work in the same way.

16. See reference made to this innovation above, p. 95.

17. A considerable amount of work was done with Dije during the early months of 1981, as she also is an accomplished potter. The choice of Dije as a primary informant was based on her enthusiasm, support, and cooperation as well as her talent. She deserves special thanks for her help and friendship during Berns' stay in the Ga'anda area.

18. Kalhar's skills have been recognized for at least the past fifteen years, as B. Rubin (1970:25) made the same observation.

19. This illustration of "new brides" shows how the first five stages of marking would have looked. See above, note 4.

20. Identical descriptions of bride's houses are included in Boyle (1913) and Meek (1931 II:386). The importance of maintaining this architectural tradition continues; grass panels are still tied over the mud walls of modern compounds to provide a suitable armature for these decorative attachments.

21. See the passage cited above (Chapter 2, p. 63) on the position of women in Ga'anda society.

22. This bracelet may have derived from a type of spiked armlet or "knuckleduster" used elsewhere as a weapon (Wente-Lukas 1977a:177). That this bracelet is an ornamental version of a martial prototype is supported by linguistic evidence—the verb stem *rɔ?*, from which the noun derives, means "to hit in the head" (R. Newman 1971).

23. The meaning of these textures in a more literal sense and an explanation of the iconographic referents that distinguish each of these sacred vessel types has been discussed in some detail in Berns (1986a).

24. It was suggested earlier that red has important ritual and symbolic connotations for the Ga'anda (see p. 66). The fact that red hematite (*mesaktariya*) is a substance that activates spirit intervention suggests that, when applied as a ritual cosmetic on the occasion of a girl's transformation to adulthood, she is joined with the realm of the spirits. Bohannan (1956:117) makes the same point in his discussion of how the Tiv rub their skin with camwood in order to make it "glow."

25. A discussion of the internal and external factors that have influenced the evolution of Ga'anda arts is a major topic of Berns' dissertation (1986a), whose larger aim is to explore what is known and what can be reconstructed of the art and history of the Lower Gongola Basin. Much of what will be schematically presented in the following discussion has been drawn from its lengthy chapters on some of the groups represented in this collection—the Ga'anda, Gbinna, Yungur, Hona, Tera, and Dera.

·5·

DECORATED GOURDS AND HISTORY

1. See Hunwick (1976), Palmer (1926), and Smith (1976).

2. Low gives some attention to Bole history in his discussion of the Gombe emirate (1972). However, he (1972:82) indicates that the best sources for their early history are the unpublished notes prepared by two gifted colonial officials, T.F. Carlyle and R.C. Abraham, now housed in the National Archives, Kaduna (NAK). A brief section on the Bole of Fika also is included in Meek (1931 II:288–310). Additionally, John Lavers of Ahmadu Bello University, Kano, is the first professional historian to gather and analyze the traditions of the Bole of Fika. The northern Bole sphere of authority appears to have been separate from that of the southern Bole, which was located north of Gombe in the Middle Gongola around the towns of Kalam, Dukku, Gadam, and Bojude.

3. Davies offers a detailed treatment of the history of Biu and the emergence of the Pabir (1954–1956; see Appendix V). A. Rubin (1974) also has published notes on the bronze and iron objects linked with Yamta-ra-Wala that are preserved in the Pabir towns of Biu and Mandaragarau.

4. A number of books have been published on the Fulani *jihads* and the Sokoto Caliphate. For details on their impact in northeastern Nigeria, see Kirk-Greene (1969) and Low (1972); for a more general treatment, see Last (1967; 1974) and Adeleye (1971; 1974).

5. For a study of the process of "Hausaization" in northern Nigeria, see Ahdamu (1978).

6. The ethnographic richness of the northeast was noted by Kirk-Greene who regarded even modern Adamawa as an "unparalleled field of study for the anthropologist" (1969:15). Newman and Davidson who recorded the music of five groups—the Bura, the Ga'anda, the Bole, the Tera, and the Tangale—stress that these peoples have retained their ethnic character despite centuries of political and cultural domination (1971:1).

7. The value of linguistics for historical reconstruction has been defended by many African anthropologists and archaeologists, as well as linguists and historians, e.g. Murdock (1959:12), Shaw (1978:95), David (1976:238), Ehret and Posnansky (1982), Greenberg (1964), Newman (1969–1970), and Wescott (1967:54).

8. All of the groups included in this discussion, with the exception of the Pidlimndi who speak a Tera dialect, are included in Temple's (1919) and/or Meek's (1931) ethnographic compilations. Readers are directed to these sources for more detailed information. However, it should be noted that the amount of data varies considerably on each subregion, which will be reflected in the length and depth of the historical summaries provided below. Additionally, only very brief references, if any, are made to categories of artistic production in the literature. Unless otherwise indicated, the data presented on the arts have been drawn from Berns' fieldwork in the Lower Gongola Valley.

9. Ballard's linguistic model for understanding the historical, though undated, movements of people in the Middle Belt also applies here, as well as to other comparably complex linguistic areas, and its basic premise merits citation:

> Where a group is assumed to have migrated in a particular direction, we find that there are peoples of another linguistic group scattered on the periphery or otherwise divided by them. In each case these separ-

ated or peripheral groups have a much higher degree of internal linguistic differentiation than the intervenors, and we can draw a further inference that the divided groups were prior inhabitants of the general area (1971:295–296).

10. A Ga'anda story of origin is included in Hammandikko (1980:1). Numerous corroborating accounts were collected by Berns in 1980–1981 among Ga'anda subgroups. Berns also collected oral traditions among the Gbinna and the Yungur during the same season. See also Meek (1931 II:370,434).

11. Meek (1931 II:369–389,434–471) notes the points of similarity between these groups in his chapters on "The Gabin [Ga'anda]" and "The Yungur-speaking Peoples." Details on Ga'anda art and ethnography are provided above in Chapter 4.

12. Temple notes that groups united under the strongest leader in times of war and communicated by means of a *lingua franca* (1919:256,391). It may have been the considerable religious authority exercised by the Ga'anda rainmaker that was responsible for integrating the disparate peoples living within his sphere of influence.

13. Other contexts of similarity between the Ga'anda and the Yungur also were noted in the earlier chapter on gourd use: they both use wickerwork baskets to display bridewealths; they both make masks with a calabash armature for boys' initiations; and they both play gourd xylophones during funeral ceremonies.

14. No pressure-engraved gourds made by the Yungur are included in this collection. The work of female artists living in the two major centers of Diterra and Suktu were studied by Chappel (1977:46–48; figs. 83–90). On the basis of his illustrations, technical and stylistic parallels can be drawn with the Gbinna examples included in this collection.

15. More on the history of the Bata is included in the section on the Benue-Gongola Valley. See also Chappel (1977:2–4).

16. Chappel writes that the Ga'anda, like the Bata, claim to have learned how to dye their gourds red from the settled Fulani (1977:48). While Berns' Ga'anda informants did not attribute this practice to the Fulani, it is clear that the technique and the ingredients used are identical. However, it is curious that only the Ga'anda and not the Gbinna or the Yungur prefer this mode of decorative enhancement.

17. The Dera are generally known in the literature as the Kanakuru (e.g., Temple 1919; Meek II 1931; Davies 1954–1956). The name apparently derived from their morning greeting, "*Kanakuru, kanakadiga*," and still seems to be favored by much of the population over the less popular "Dera." Although Newman (1977:4) retains "Kanakuru" in his Chadic classification in order to avoid multiplying new and unfamiliar designations, Dera is used here in order to maintain standards of historical accuracy.

18. Dera oral traditions are recounted in Nissen (1968:90), Davies (1954–1956:33) and Meek (1931 II:311). Berns also collected complementary accounts from elders living in the chiefdoms of Shani, Shellen, and Kiri in 1980–1981.

19. Ga'anda oral traditions specify that a third section of their migratory group (the Ga'anda and Hona being the first and the second) were cattle herders who continued further west in search of better pasturage (Hammandikko 1980:1). They eventually settled in the hills flanking the Gongola flood plain around Shani.

20. Newman (1969–1970:219). Temple also cites divergent Tera migrations, with one group coming from Fika and the other from Shani (1919:350–351).

21. This same phenomenon may explain the distinct historical traditions maintained by Dera villages outlined above. There, however, it may have been Tera- and/or Yungur-speakers who adopted a Dera language in place of their own. The propensity for this phenomenon among Gongola-Hawal peoples and those of the Ga'anda Hills as well, points up the risks in uncritically drawing inferences about common origins strictly from present linguistic affinities. The role of art in understanding such complex historical relationships is treated in greater detail in Berns (1986a).

22. Information about traditional Dera practices is included in Meek (1931 II:311–325). Unpublished notes also are available in the National Archives, Kaduna (NAK): files NAK 2710C, 2844.

23. For other illustrations of painted Dera (Kanakuru) vessels, see Leith-Ross (1970:45–48).

24. Cases of "language shifting" also have been documented between the geographically proximate Tera and Bura. Newman reports that one major Bura town that lies at their frontier, called Kwaya Kusar, was previously Tera-speaking (1969–1970: 221). In this instance, the Tera dropped their own language and adopted that of the more dominant Bura, who moved into this area from their hill outposts on the adjacent Biu Plateau.

25. Berns worked briefly in two southern Bole towns—Kwami (formerly the Tera village of Kwamu, later absorbed by the large Bole town of Kafarati, and more recently renamed Kwami) and Ribadu (an old and historically important Bole town)—where the tradition of women's gourd decoration has largely been abandoned. She photographed one Bole gourd "sun bonnet" (*koliku*) that was engraved identically—both inside and out—to those made by Tera and Bura artists (see Fig. 24). In 1981–1982, even the gourds surmounting the pottery vessels lining the walls of Bole women's rooms were left undecorated. However, no gourds decorated by the southern Bole are included in this collection, nor does this group appear to have developed as elaborate a tradition as that of their Tera neighbors.

26. One Bole pot-stand, collected in the northern capital of Fika by Olive Macleod, is now housed in the Museum of Mankind, London [1911.12–14.17]. A very similar example is illustrated in Leith-Ross (1970:39). Although somewhat different in profile, clearly related examples called *almari*, were documented in 1982 by Berns in the southern Bole town of Ribadu.

27. In addition to Meek's (1931:I and II) sketches of each group's distinctive facial markings, Bole striations are described by Temple (1919:63) and Merrick (1904–1905:418).

28. Temple describes Kanuri tribal marks as follows: ". . . variously ten to twelve parallel lines from the temples to the level of the mouth (Bauchi); seven lines on the right cheek with six lines above; ten lines of the left cheek with six lines above,

and one zara (Sokoto)" (1919:221). She also writes that Bole marks "are possibly an exaggerated form of Kanuri" (1919:63). For information on the early history of the Kanuri in the Lake Chad Basin, see Smith (1976:158–171) and Hunwick (1976: 264–274); Temple (1919:217–219) also includes a few historical passages drawn from early travelers' accounts.

29. This story of Bole migration is reported in Newman (1969–1970:218) and is corroborated by Temple (1919:62), who based her account on the "Fikan chronicles."

30. The likelihood that the early occupants of the Chad Basin practiced similar patterns of facial scarification is suggested by their depiction in more permanent sculptural materials. For example, one of the small, relatively naturalistic, terracotta heads in the "Sao" archaeological corpus is covered with dense linear incisions; other small "millstone-shaped" terracotta heads (ca. 10 cm) are covered with linear incisions that may more schematically depict the same custom (A.M.D. and J.-P. Lebeuf 1977:figs. 69; 37–46). While there is some controversy about their chronology, the Lebeufs date the Sao terracottas from the tenth to the thirteenth centuries A.D. (1977:194–195). The patterns of facial striations known from northeastern Nigeria also have been compared to the vertical scoring that distinguishes Ife brass and terracotta heads (e.g., Willet 1967:fig. 33; Newman and Davidson 1971:7). Where references in the literature tend to focus specifically on the Tera, the information presented here suggests that this tradition was far more widespread in northeastern Nigeria. That this pattern of scarification is also evident in the Lake Chad Basin—long a major trading entrepôt and center of considerable political power—has special implications for the study of Ife material. For example, the likelihood that the ancient Ife kingdom engaged in intensive commerical transactions with northern traders in order to import copper or brass (as well as other prestige materials), contributes to the plausibility of such long-range artistic parallels.

31. Low identifies the Fulani clans who settled in the Lower Gongola Basin as the Kitaku or Kitiyen (1972:74; n. 47). He also notes the role of each settlement during the nineteenth-century *jihads* (1972:86–89). Leaders of these communities later became the governing elite of the Gombe Emirate. See Low's map for the geographic distribution of these Kitaku settlements (1972:83).

32. It should be noted that the languages spoken in this subregion are particularly well-documented, making their classification of special value for understanding historical relationships (pc:R. Schuh 1984).

33. The history of the northern Bole dominates what is known about the region. For references about Bole history, see note 2.

34. It should be noted that there is no historical or even stylistic basis for suggesting a connection between the Potiskum method of gourd painting and the way the settled Fulani paint the interior of their gourds. Whereas Fulani artists practicing this technique are relatively rare, most Karekare, Ngamo, and Ngizim women decorate their gourds in this fashion.

35. This information is based on Hudson's interview in January 1971 with one male Bole carver, named Dambaya, who was living in Fika. He had learned pyro-engraving from a man living in another Bole village and at the time of the interview was teaching young boys this technique in Fika.

36. Temple reports that Kilba men and women were both "extensively tattooed" (1919:230). Meek (1931 I:202) confirms this report and indicates that, while no longer a strict premarital requirement, Kilba girls are usually cicatrized before going to their husbands. In addition to Meek's (1931 I:233) sketch of typical Margi facial striations, he illustrates "Margi maidens" whose abdomens are covered with rows of short, linear cicatrices (facing 214). Vaughan mentions that eastern Margi blacksmiths are responsible for scarifications, although he does not describe what they look like (1975:167). The fact that other Mandara groups practice scarification that takes the form of cicatrices rather than recessed "cuts," agrees with the proposal made earlier that their respective distribution may be related to the migrations of Biu-Mandara versus West Chadic peoples. More on this subject is included in Berns (1986a).

37. Vaughan has analyzed in some detail the Margi tradition of cultural and political independence from their Nigerian neighbors, particularly the Fulani (1964; 1981). He notes, however, that after the second plebiscite in 1961 when the Margi voted to join the Federation of Nigeria, they reversed their position and converted to Islam in order to remain in step with groups of higher authority in the northeast. This resolution of past antagonisms, especially to Fulani incorporation, reflects the Margi capacity to make changes furthering their own self-interests. According to this thesis, the Margi adoption of pyro-engraving before 1961 suggests that while symbolically or technically advantageous, it was not threatening to social or cultural autonomy.

38. For a discussion of Margi blacksmiths, see Vaughan (1975). For related Mandara traditions, see also Podlewski (1966), Sassoon (1964), and Wente-Lukas (1977b).

39. A. Rubin (1974) has shown the ways in which the bronze and iron regalia associated with Pabir royalty has important implications for our understanding of the history of northeastern Nigeria. Specifically, the material evidence he presents argues against the assumption that a pre-Pabir Jukun stratum once dominated the Biu area. The same information also may be used to link the Bura and the Pabir with related peoples living to the east in the Uba Plains, as well as to the west and to the south in the Gongola-Hawal Valley.

40. See Wente-Lukas (1977a:41). There is evidence that women living in the Matakam village of Dgrwada ornament gourds using a similar process to that of the Margi. Also, gourds photographed among the Fali of Vimtim by Berns in 1982 were decorated using resist-dyeing.

41. Ballard's linguistic model also applies here—the Bata have clearly migrated away from the area where their closest linguistic relatives are now dispersed, to an area dominated by Adamawa speakers (see above, note 9). David (1976:242) agrees that the Bata are a Chadic intrusion into an Adamawa-speaking landscape.

42. This generalization has been made based on the oral accounts collected by Berns in 1982 among a number of Adamawa-speaking peoples living northwest of the confluence. Similarly, Yungur-speakers refer to a local hill site (Mukan) as their ancient homeland.

43. Although the historical notes about this region provided

by Kirk-Greene (1969:17) and Stevens (1976:31) confirm that the Bata invaded and conquered the Upper Benue, Stevens also indicates that there is some confusion as to the pre-Fulani (i.e., prenineteenth century) history of the Chamba. However, on the basis of linguistic geography, it is likely that the Chamba were displaced by an intrusive population.

44. This important cult is described in detail in Meek (1931 I:25–44) and in Kirk-Greene (1969:appendix C). Stevens (1976: 31–32) describes how Nzeanzo also may provide an historical link between the Bachama and the Chamba.

45. See Chapter 2, p. 66. Very little ethnographic work has been done on the Mbula who tend to be treated as an off-shoot of the Bata-Bachama cluster. Rather than having a "derivative" culture, the Mbula may have had more impact on surrounding populations than is otherwise thought. It is hoped that the data Berns collected on the Mbula in 1982, with a special emphasis on their arts and ritual practices, together with the information that has been compiled by Mr. Yustas Offah (n.d.) on their history and traditions, will yield new insights into the historical and cultural position of this group.

46. Hudson photographed one calabash in the town of Numan in 1971 made by a Bachama woman. Although a single example, it does illustrate the stylistic features that are shared by the Mbula, the Bachama, and the Bata of Geren.

47. The association of figurated ceramic sculpture with Adamawa peoples extends to groups living south of the Benue as well. Although this point cannot be developed or further substantiated here, it is discussed in Berns (1986a). Articles published on ceramic arts from the Lower Gongola Valley are listed in Chapter 1, note 22. Additionally, a publication focusing on the distribution of ceramic sculpture in this region is anticipated by Berns for the future.

·6·

GOURDS AND MODERN CHANGE

Tradition vs. New Directions

1. In 1981, Berns was invited to attend a wedding celebration in the small village of Ganjara located on the dry season road linking Ga'anda to the town of Gombi. The bride had displayed in her parents' compound an incredible array of enamel containers, pyrex casseroles, cutlery, crockery, glassware, and aluminum cookware, as well as kerosene stoves and lamps, printed cloth, blankets, and ready-made clothing.

2. In response to Bravmann's (1974) theoretical premises, Chappel (1977:176) has argued that the syncretic nature of Islam among the Hausa, for example, is not evident among the Adamawa Fulani. Rather than Islam accommodating itself to local circumstances in order to attract converts, people have completely abandoned their traditions to elevate their status vis-a-vis that of the "imperial" Fulani. This has had a deleterious effect on the arts associated with religious practices, including symbols of cult membership and objects used in ritual worship (see Chappel 1973). Although today there is little overt opposition to individuals who have chosen to maintain traditional practices, the incorporation of gourds in secret societies, burials, funerals, sacred libations, and divination procedures probably has a limited future based on present rates of religious conversion.

3. Even in the early 1970s, Hudson was presented with gifts of enamelware dishes and bowls despite her clear research interests.

4. There is a preference for traditional customs like gourd decoration even though groups may regard the gourds themselves as outmoded. Chappel (1977:178–180) writes that the Yungur attitude about gourd decoration was like that held about language—it was something that ought to be preserved in spite of Christian, Islamic, or Western influence. It is this kind of "cultural chauvinism" that may provide the impetus for maintaining such secular artistic traditions.

5. Unfortunately, no information on the provenance of this gourd was collected from the Waja woman who was demonstrating pyro-engraving at the display staged in Dela Waja village (1982). It resembles, however, two examples reproduced in Chappel made by male Jafun, as well as Daneji Fulani, carvers (1977:figs. 63,166). They, too, have unusual and complex pictorial images that Chappel associates with the fairly free style of pastoral Fulani men (1977:178). He claims that they have openly taken up this art because they find it personally challenging as well as commercially rewarding. They also enjoy a certain measure of freedom in constructing compositions because religious restraints do not prevent them from using representational imagery.

6. Information on this modern decorative tradition is based on an interview conducted by Berns with Mohammadu Gombe in Dumne market, June 1981.

7. It is likely that openwork carving was borrowed from the Yoruba (southwestern Nigeria) who have a tradition of carving thick-walled gourds with pierced shapes, suitable for holding solid objects (Ojo 1966:248; pl. 25). In the last two decades, this mode of carving also has become a major tourist art in northern as well as southern Nigeria.

8. It should be noted that the National Museum, Jos (formerly the Federal Department of Antiquities) has a duplicate collection of decorated gourds assembled by Hudson in 1970–1971. Chappel also collected a substantial number of gourds for the Museum in 1965–1966.

APPENDIX

1. This list has been extracted from Chappel (1977:36–40) in addition to Taylor (1932).

2. Many of the Hausa terms were included in Heathcote's (1976:96–100) useful English-Hausa vocabulary of art related terms. His entries were checked against and supplemented by Abraham (1962) and Dalziel (1916:26–27). The following orthographic symbols for Hausa and other Chadic languages should be noted:

b' = glottalized "b"
d' = glottalized "d"
k' = glottalized "k"

3. Ngizim terms have been drawn from Schuh (1981).

4. These words, field collected by Berns, were checked against lists provided in Newman (1974).

5. Ga'anda terms were field collected by Berns and checked for accuracy when possible against R. Newman (1971). The following additional orthographic conventions for Ga'anda should be noted:

x = voiceless velar fricative
ɇ = schwa
? = glottal stop
hl = voiceless lateral fricative
n' = voiced velar nasal

6. The list of Hona words as well as those of the remaining groups were field collected and transcribed by Berns to the best of her ability.

7. One other orthographic convention should be noted for Gbinna, Yungur, and Mumuye: ~ = nasalized vowel.

BIBLIOGRAPHY

Abraham, R.C.
1962 [1946] *Dictionary of the Hausa Language*. London.

Adeleye, R.A.
1971 *Power and Diplomacy in Northern Nigeria 1804–1906. The Sokoto Caliphate and its Enemies*. New York.
1974 "The Sokoto Caliphate in the Nineteenth Century." In *History of West Africa*. J.F.A. Ajayi and M. Crowder (eds.). Vol. 2. New York.

Ahdamu, M.
1978 *The Hausa Factor in West African History*. Ahmadu Bello University History Series. Zaria, Nigeria.

Aitchison, P.J., M.G. Bawden, D.M. Carroll, P.E. Glover, K. Klinkenberg, P.N. de Leeuw, and P. Tuley
1972 *Land Resources of North East Nigeria*. Vol. 1, The Environment. Land Resource Study No. 9. Foreign and Commonwealth Office, Overseas Development Administration. Surrey, England.

Ajayi, J.F.A. and M. Crowder (eds.)
1974 *History of West Africa*. 2 vols. New York.

Ames, D.W.
1968–1969 "Anthology of African Music: The Music of Nigeria, Hausa Music." UNESCO Collection. 2 vols. Barenreiter Musicaphon.

Ames, D.W. and A.V. King
1971 *Glossary of Hausa Music and its Social Contexts*. Evanston, Illinois.

Anonymous
1924 "European and African Workers for Wembley." *West Africa* (Empire Exhibition Supplement).

Arnheim, R.
1967 *Art and Visual Perception: A Psychology of the Creative Eye*. London.

Arriens, C.
1928 *Am Herdfeuer der Schwarzen: Erlebtes aus Westafrika*. Weimar.

Astley, T. (ed.)
1745–1747 *A new general collection of voyages and travels consisting of the most esteemed relations, which have been hitherto published in many languages: comprehending everything remarkable in its kind, in Europe, Asia, Africa and America*. 4 vols. London.

Atkins, G. (ed.)
1972 *Manding Art and Civilization*. London.

Balandier, G. and J. Maquet (eds.)
1974 *Dictionary of Black African Civilization*. New York.

Ballard, J.A.
1971 "Historical inferences from the linguistic geography of the Nigerian Middle Belt." *Africa* 41:294–305.

Beckwith, C. and M. van Offelen
1983 *Nomads of Niger*. New York.

Berns, M.
1978a Review of *Unearthing Igbo Ukwu . . .* by T. Shaw. *African Arts* 11(4):14–19.
1978b Review of *Les arts des Sao* by J.-P. and A.M.D. Lebeuf. *African Arts* 12(1):103–106.
1985 "Decorated Gourds of Northeastern Nigeria." *African Arts* 19(1):28–44,86–87.
1986a "Art and History in the Lower Gongola Basin, Northeastern Nigeria." Ph.D. dissertation, University of California, Los Angeles.
1986b "Ga'anda Scarification: A Model for Art and Identity." In *Anthropometamorphosis*. A. Rubin (ed.). Los Angeles. Forthcoming.

Boahen, A. et al.
1971 *The Horizon History of Africa*. New York.

Bohannan, P.
1956 "Beauty and Scarification Amongst the Tiv." *Man* 56(129):117–121.

Bossert, H.T.
1955 *Folk Art of Primitive People*. New York.

Boyle, C.V.
1913 "Lala District: Historical and Anthropo-

logical Notes." In *NAK* [J.18]. 1912–1928: 12–38.
1916a "The Ordeal of Manhood." *Journal of the African Society* 15(59):244–255.
1916b "The Marking of Girls at Ga'anda." *Journal of the African Society* 15(60):361–366.

Bravmann, R.
1973 "Open Frontiers: The Mobility of Art in Black Africa." *Index of Art in the Pacific Northwest*, no. 5. Seattle.
1974 *Islam and Tribal Art in West Africa*. London.

Brincard, M.-T. (ed.)
1982 *The Art of Metal in Africa*. New York.

Burton, R. F.
1869 *The Lake Regions of Central Africa*. New York.

Butler, J. (ed.)
1964 *Boston University Papers in African History*. Boston.

Chappel, T.J.H.
1973 "The Death of a Cult in Northern Nigeria." *African Arts* 6(4):70–74.
1977 *Decorated Gourds in North-Eastern Nigeria*. London.

Cole, H.M. (ed.)
1985 *I Am Not Myself: The Art of African Masquerade*. Los Angeles.

Cole, H.M. and C. Aniakor
1984 *Igbo Arts: Community and Cosmos*. Los Angeles.

Connah, G.
1976 "The Daima Sequence and the Prehistoric Chronology of the Lake Chad Region of Nigeria." *Journal of African History* 18(3): 321–352.

Coppens, M.
n.d. *Sculpturen van Noordoost Nigeria. Bauchi, Waja, Wurkum*. Eindhoven, Netherlands.

Dalby, W.
1972 "Music." In *Manding Art and Civilization*. G. Atkins (ed.). London.

Dalziel, J.M.
1916 *A Hausa Botanical Vocabulary*. London.

David, N.
1976 "History of Crops and Peoples in North Cameroon to A.D. 1900." In *Origins of African Plant Domestication*. J.R. Harlan, J.M.J. DeWet, and B.L. Stemler (eds.). The Hague and Chicago.

David, N. and H. Hennig
1972 *The Ethnography of Pottery: A Fulani Case Seen in Archaeological Perspective*. McCaleb Module in Anthropology from Addison-Wesley Modular Publications, no. 21:1–29.

Davies, J.G.
1954–1956 *The Biu Book. A Collation and Reference Book on the Biu Division*. Zaria, Nigeria.

d'Azevedo, W.L. (ed.)
1975 *The Traditional Artist in African Societies*. Bloomington, Indiana.

Dodge, E.
1943 *Gourd Growers of the South Seas: An Introduction to the Study of the Lagenaria Gourd in the Culture of the Polynesians*. The Gourd Society of Ameria. Ethnographical Series No. 2.

Dunhill, A.
1969 *The Pipe Book*. London.

Dupire, M.
1963 "The position of women in a pastoral society." In *Women of Tropical Africa*. D. Paulme (ed.). London.

Ehret, C. and M. Posnansky
1982 *The Archaeological and Linguistic Reconstruction of African History*. Berkeley and Los Angeles.

Fagg, W.
1972 *The Living Arts of Nigeria*. New York.

Frobenius, L.
1923 *Das sterbende Afrika*. Vol. 1. Munich.

Gabel, C. and N. Bennett (eds.)
1967 *Reconstructing African Culture History*. Boston.

Gabus, J.
1967 *Art negre: Recherche de ses fonctions et dimensions*. Neuchatel.

Gardi, R.
1969 *African Crafts and Craftsmen*. New York.

Gauthier, J.-G. and G. Jansen
1973 *Ancient Art of the Northern Cameroons: Sao and Fali*. Anthropological Publications. Oosterhout, Netherlands.

Greenberg, J.
1964 "Historical inferences from linguistic research." In *Boston University Papers in African History*. J. Butler (ed.) Boston.
1966 *The Languages of Africa*. Bloomington, Indiana.

Griaule, M. and G. Dieterlen
1935 "Calebasses dahoméennes." *Journal de la société des Africanistes* 5(3):203–207.

Hambly, W.D.
1935 *Culture Areas of Nigeria*. Field Museum of Natural History, Chicago. Anthropological Series 21(3).

Hamelin, P.
1952–1953 "Les bronzes du Tchad." *Tribus*:379–399.

Hammandikko, M.
1980 "History of Ga'anda/Tarihin Ga'anda." Edited and translated by M. Berns. *Occasional Paper No. 21*, African Studies Center, University of California, Los Angeles.

Hansford, K., J. Bendor-Samuel, and R. Stanford
1976a "An Index of Nigerian Languages." *Studies in Nigerian Languages*, no. 5.
1976b "A Provisional Language Map of Nigeria." *Savanna* 5(2):115–126.

Harlan, J.R., J.M.J. DeWet, and B.L. Stemler (eds.)
1976 *Origins of African Plant Domestication*. The Hague and Chicago.

Harvey, C.
1985 *Austin Seven*. Somerset, England.

Heathcote, D.
1976 *The Arts of the Hausa*. London.

Heiser, C.
1979 *The Gourd Book*. Norman, Oklahoma.

Herskovits, M.
1938 *Dahomey: An Ancient West African Kingdom*. New York.

Hodge, A.
1982 *Nigeria's Traditional Crafts*. London.

Huet, M.
1954 *Les hommes de la danse*. Lausanne.

Huet, M., J. Laude, and J.-L. Paudrat
1978 *The Dance, Art and Ritual of Africa*. New York.

Hunwick, J.
1976 "Songhay, Borno and Hausaland in the sixteenth century." In *History of West Africa*. J.F.A. Ajayi and M. Crowder (eds.). New York.

Isichei, E. (ed.)
1982 *Studies in the History of the Plateau State*. London and Basingstoke.

Jefferson, L.E.
1973 *The Decorative Arts of Africa*. New York.

Jobson, R.
1623 [1968] *The Golden Trade, or a Discovery of the River Gambra and the Golden Trade of the Aethiopians*. London.

Josephy, A.M. Jr. (ed.)
1971 *The Horizon History of Africa*. New York.

Junod, H.-P.
1936 "The Vachopi of Portuguese East Africa." *Bantu Tribes of South Africa* 4(2).
1938 *Bantu Heritage*. Johannesburg.

Kandert, J.
1974 "Folk Pottery of Nigeria." *Annals of the Náprstek Museum*, no. 7.

Kirby, P.R.
1934 *The Musical Instruments of the Native Races of South Africa*. London.

Kirby, R.
1984 *Gourds and Calabashes*. London.

Kirk-Greene, A.H.M.
1969[1958] *Adamawa Past and Present. An historical approach to the development of a Northern Cameroons Province*. Oxford.

Konan, M.
1974 "Calabashes in Northern Nigeria." *Expedition* 17(1):2–10.

Krieger, K.
1965–1969 *Westafrikanische Plastik*. 3 vols. Berlin.
1968 "Musikinstrumente der Hausa." *Baessler-Archiv*, Neue Folge, 16:373–429.

Last, D.M.
1967 *The Sokoto Caliphate*. London.
1974 "Reform in West Africa: The Jihad Movements of the Nineteenth Century." In *History of West Africa*. J.F.A. Ajayi and M. Crowder (eds.). Vol. 2. New York.

Leakey, M.D. and L.S.B. Leakey
1950 *Excavations at the Njoro River Cave*. London.

Lebeuf, J.-P.
1982 "The ancient metallurgy of copper and its alloys in the region of Lake Chad." In *The Art of Metal in Africa*. M.-T. Brincard (ed.). New York.

Lebeuf, J.-P. and A.M.D. Lebeuf
1977 *Les Arts des Sao*. Paris.

Lebeuf, J.-P. and J. Françaix
1978 "Analyse d'objets de métal en provenance du Cameroun et du Tchad." *Objets et Mondes* 18(1–2):96–100.

Leith-Ross, S.
1970 *Nigerian Pottery*. Ibadan, Nigeria.

Low, V.
1972 *Three Nigerian Emirates. A Study in Oral History*. Evanston, Illinois.

Mackay, M.
1955 "The *shantu* music of the *harims* of Nigeria." *African Music* 1(2):56–57.

Macleod, O.
1912 *Chiefs and Cities of Central Africa. Across Lake Chad by way of British, French and German Territories*. Edinburgh and London.

Malzy, P.
1957 "Les Calebasses." *Notes Africaines* 73:10–12.

Mauk, P.A.
1980 "Adamawan Calabashes and a Pastoral Fulani Aesthetic." Unpublished M.A. thesis. University of Washington, Seattle.

Meek, C.K.
1925 *The Northern Tribes of Nigeria*. 2 vols. London.
1931 *Tribal Studies in Northern Nigeria*. 2 vols. London.

Mercier, P.
1974 "Calabashes." In *Dictionary of Black African Civilization*. G. Balandier and J. Maquet (eds.). New York.

Merrick, G.
1904–1905 "The Bolewa Tribe." *Journal of the African Society* 4:417–426.

Mordecai, G.
1978 *Gourd Craft*. New York.

Murdock, G.
1959 *Africa. Its Peoples and their Culture History*. New York.

Murray, K.C.
1951 "The decoration of Calabashes by Tiv." *Nigeria* 36:469–473.

National Archives, Kaduna (NAK)
Adamawa Provincial Archives
1912–1928 [J.18]. "The Lala and Hona. Collected Historical and Anthropological Papers."
Secretariat of the Northern Provinces
1924–1927 [2710C]. "Ethnology. Kanakuru Tribe (Shellem)." D.F. Heath, "Kanakuru initiation ceremony." [1927:no. P.E./K./4].
1912–1919 [2844]. "Assessment Reports on the Kanakuru Districts." W.O.P. Rosedale, "Kanakuru Assessment Report." [1919].
1914 [10/21179]. T.F. Carlyle, "History of the Gombe Emirate, Central Province."
1926 [17/8 K1119]. R.C. Abraham, "Ethnological Notes on the Bolewa Group."

Neher, G.
1964 "Brass Casting in North-East Nigeria." *Nigerian Field* 29(1):16–27.

Neaher, N.
1979 "Nigerian Brass Bells." *African Arts* 12(3): 42–47,95.

Newman, P.
1969–1970 "Linguistic relationship, language shifting and historical inference." *Afrika und Ubersee* 53:217–223.
1974 *The Kanakuru Language*. West African Language Monograph Series, no. 9.
1977 "Chadic Classification and Reconstructions." *Monographic Journals of the Near East. Afroasiatic Linguistics*. December 5(1).

Newman, P. and E. Davidson
1971 "Music from the villages of Northeastern

Nigeria." *ASCH Mankind Series*, Album No. AHM 4532.

Newman, R.
1971 "A Case Grammar of Ga'anda." Ph.D. dissertation, University of California, Los Angeles.

Newman, T.R.
1974 *Contemporary African Arts and Crafts*. New York.

Nicklin, K.
1982 "The Cross River Bronzes." In *The Art of Metal in Africa*. M.-T. Brincard (ed.). New York.

Nishimura, S.
1975 "Styles of Ornament on Calabashes in West Africa." *Kyoto University African Studies* 9: 91–103.

Nissen, M.
1968 *An African Church is Born: The Story of the Adamawa and Central Sardauna Provinces in Nigeria*. Denmark.

Northern, T.
1975 *The Sign of the Leopard: Beaded Art of Cameroon*. Storrs, Connecticut.
1984 *The Art of Cameroon*. Washington, D.C.

Offah, Y.
n.d. "Tarihin Mbula/History of Mbula." Cyclostyled. Zing Local Government. Gongola State, Nigeria.

Ojo, G.J.A.
1966 *Yoruba Culture: a geographical analysis*. London.

Palmer, H.R. (trans.)
1926 *History of the First Twelve Years of the Reign of Mai Idris Alooma of Bornu (1571–1583)* by his Imam Ahmed ibn Fartua. Lagos.

Parrinder, G.
1967 *African Mythology*. London.

Paulme, D.
1963 *Women of Tropical Africa*. London.

Paulme-Schaeffner, D.
1949 "Pottery Grave and House Ornaments in West Africa." *Man* 49:24.

Pearlstone, Z.
1973 "Sujang: A Stirrup Spout Vessel from Nigeria." *American Antiquity* 38(4):482–486.

Perani, J.
1977 Review of *Decorated Gourds in North-Eastern Nigeria* by T.J.H. Chappel. *African Arts* 11(1): 9–10.
1985 "Aspects of Creativity in Hausa Calabash Decoration." Paper presented at Male and Female Artistry Symposium, April 25, 1985. University of Minnesota.

Pevar, S.G.
1978 "The Construction of a Kora." *African Arts* 11(4):66–72.

Pharr, L.
1983 "The Ubiquitous Gourd: The Art of Calabash Carving in Africa." Photocopied annotated bibliography. J. Stanley (ed.). Smithsonian Institution Libraries. Washington, D.C.

Podlewski, A.M.
1966 "Les forgerons Mafa: Description et évolution d'un groupe endogame." *Cahiers O.R.S.T.O.M.* Série Sciences Humaines 3(1).

Price, C.
1975 *Made in West Africa*. New York.

Price, S.
1982 "When is a Calabash Not a Calabash?" *New West Indian Guide* 56(1/2):69–82.

Roscoe, J.
1915 *The Northern Bantu*. Cambridge.

Rubin, A.
1973 "Bronzes of the Middle Benue." *West African Journal of Archaeology* 3:221–231.
1974 "Notes on Regalia in Biu Division, North-Eastern State, Nigeria." *West African Journal of Archaeology* 4:161–175.
1976 "Styles of African Sculpture." In *The Sculptor's Eye: The African Art Collection of Mr. and Mrs. Chaim Gross*. Museum of African Art, Washington, D.C.
1982 "Prologue to Art History in Plateau State." In *Studies in the History of the Plateau State*. E. Isichei (ed.). London and Basingstoke.
1985 "A Mumuye Mask." In *I Am Not Myself: The Art of African Masquerade*. H.M. Cole (ed.). Los Angeles.
1986 *Anthropometamorphosis* (ed.). Los Angeles.
Forthcoming *Sculpture of the Benue River Valley*. Los Angeles.

Rubin, B.
1970 "Calabash Decoration in North East State, Nigeria." *African Arts* 4(1):20–25.

Saitoti, T. and C. Beckwith
1980 *Maasai*. New York.

Sassoon, H.
1964 "Iron smelting in the hill village of Sukur, North-Eastern Nigeria." *Man* 215:174–178.

Schuh, R.
1981 *A Dictionary of Ngizim*. Los Angeles and Berkeley.

Shaw, T.
1969 "Archaeology in Nigeria." *Antiquity* 43: 187–199.
1970 *Igbo Ukwu. An account of archaeological discoveries in Eastern Nigeria*. 2 vols. Great Britain.
1977 *Unearthing Igbo Ukwu. Archaeological discoveries in Eastern Nigeria*. Ibadan.
1978 *Nigeria. Its Archaeology and Early History*. London.

Sieber, R.
1980 *African Furniture and Household Objects*. Bloomington, Indiana.

Sieber, R. and T. Vevers
1974 *Interaction: The art styles of the Benue River Valley and East Nigeria*. West Lafayette, Indiana.

Skinner, N.
1968 *Hausa Readings. Selections from Edgar's 'Tatsuniyoyi'*. Madison, Wisconsin.

Slye, J.
1969 "Pagan Ju-Ju Ceramics in the Northern States." *Nigeria Magazine* 102:496–515.
1977 "Mwona Figurines." *African Arts* 10(4):23.

Smith, A.
1976 "The early states of the Central Sudan." In *History of West Africa*. J.F.A. Ajayi and M. Crowder (eds.). New York.

Stenning, D.
1952 — *Savannah Nomads*. London.

Stevens, P.
1976 — "The Danubi Ancestral Shrine." *African Arts* 10(1):30–37,98.

Taylor, F.W.
1932 — *A Fulani-English Dictionary*. Oxford.

Teilhet, J.
1977–1978 — "The Equivocal Role of Women in Non-Literate Cultures." *Heresies* 1(4):96–102.

Temple, C.L.
1919 — *Notes on the Tribes, Provinces, Emirates and States of the Northern Provinces of Nigeria*. London.

Trowell, M.
1960 — *African Design*. New York.

Udo, R.
1970 — *Geographical Regions of Nigeria*. Berkeley.

Vaughan, J.
1964 — "Culture, history and grass-roots politics in a northern Cameroon Kingdom." *American Anthropologist* 66(5):1078–1095.
1975 — "*en'kyagu* as artists in Marghi society." In *The Traditional Artist in African Societies*. W.L. d'Azevedo (ed.). Bloomington, Indiana.
1981 — "Margi Resistance to Fulani Incorporation: A Curious Resolution." *Image and Reality in African Interethnic Relations. The Fulbe and their Neighbors*. E. Schultz (ed.). Studies in Third World Societies, no. 11.

Weber, C.
1962 — *Weber's Guide to Pipes and Smoking*. New York.

Wente-Lukas, R.
1977a — *Die materielle Kultur der nicht-islamischen Ethnien von Nordkamerun und Nordostnigeria*. Studien zur Kulturkunde, no. 43. Wiesbaden.
1977b — "Fer et forgeron au sud du lac Tchad (Cameroun, Nigeria)." *Journal des Africanistes* 47(2): 107–122.

Westcott, R.W.
1967 — "African Languages and African Pre-history." In *Reconstructing African Culture History*. C. Gabel and N. Bennett (eds.). Boston.

Western, D.C.
1975 — *A Bibliography of the Arts of Africa*. African Studies Association, Brandeis University. Waltham, Massachusetts.

Whitting, C.E.J.
1940 — *Hausa and Fulani Proverbs*. Lagos.

Willet, F.
1967 — *Ife in the History of West African Sculpture*. New York.

Wittmer, M. and W. Arnett
1978 — *Three Rivers of Nigeria: Art of the Lower Niger, Cross and Benue from the Collection of William and Robert Arnett*. High Museum of Art, Atlanta.

Wyatt, R.J.
1968 — *The Austin Seven*. Trowbridge, England.

PRESENTATION

DARLENE MOSES OLYMPIUS *Design; Map A, B, and C*
IRINA AVERKIEFF *Editing*
RICHARD TODD *Studio Photography*
ROBERTA ROBERTS *Map D*
NANCY TOOTHMAN *Illustrations*
ELLEN HARDY *Photographic Assistance*
SUSAN SWISS *Editorial Assistance*
UCLA PUBLICATION SERVICES *Production Management*
AMERICAN TYPESETTING, INC. *Typography and Production Art*
ALAN LITHOGRAPH INC. *Color Separations and Printing*
ROSWELL BOOKBINDING *Binding*

Photography Credits: Carol Beckwith, Pl. 8 (from *Nomads of Niger*, published by Harry N. Abrams, Inc., 1983). Marla C. Berns, Pls. 1–4, 9–17, 19, 34, 36–38; Figs. 1, 6, 18, 19, 22, 29, 30–35, 38–40, 50, 53, 72, 82, 95, 105, 111, 114, 115, 117–119, 120b, 121b, 124, 125, 127, 130, 132, 136, 137, 142, 144, 145, 146, 150. T.J.H. Chappel, Fig. 12. Barbara Rubin Hudson, Frontispiece; Pl. 6; Figs. 9, 21, 23, 66, 89, 134. National Museum, Lagos, Figs. 2, 3. Merrick Posnansky, Fig. 5. Arnold Rubin, Figs. 36, 37, 41, 42, 106. Russell Schuh, Pl. 5; Fig. 147.

MUSEUM STAFF